Pragmatist Neurophilosophy

Bloomsbury Studies in American Philosophy

Bloomsbury Studies in American Philosophy presents cutting-edge scholarship in both the history of and contemporary movements in American philosophy. The wholly original arguments, perspectives and research findings in titles in this series make it an important and stimulating resource for students and academics from across the field.

America's First Women Philosophers, Dorothy G. Rogers
Dewey and the Ancients, edited by Christopher C. Kirby
Exploring the Work of Edward S. Casey, edited by Azucena Cruz-Pierre and Donald A. Landes
Feminist Epistemology and American Pragmatism, Alexandra L. Shuford
John Searle and the Construction of Social Reality, Joshua Rust
The Legacy of John Rawls, edited by Thom Brooks and Fabian Freyenhagen
Nozick, Autonomy and Compensation, Dale F. Murray
Peirce, James, and a Pragmatic Philosophy of Religion, John W. Woell
Peirce's Philosophy of Communication, Mats Bergman
Peirce's Pragmatic Theory of Inquiry, Elizabeth Cooke
Philosophy of History After Hayden White, edited by Robert Doran
Pragmatist Metaphysics, Sami Pihlström
Quine on Meaning, Eve Gaudet
Quine's Naturalism, Paul A. Gregory
Reality and Its Appearance, Nicholas Rescher
Relativism in Contemporary American Philosophy, Timothy M. Mosteller
Richard Rorty, edited by Alexander Gröschner, Colin Koopman and Mike Sandbothe
Richard Rorty's New Pragmatism, Edward J. Grippe
Thomas Kuhn's Revolution, James A. Marcum
Varieties of Pragmatism, Douglas McDermid
Virtue Ethics: Dewey and MacIntyre , Stephen Carden

Also available from Bloomsbury:

The Bloomsbury Encyclopedia of Philosophers in America, edited by John R. Shook
The Cambridge School of Pragmatism, edited by André De Tienne and John R. Shook
The Dictionary of Early American Philosophers, edited by John R. Shook
The Dictionary of Modern American Philosophers, edited by John R. Shook

Pragmatist Neurophilosophy

American Philosophy and the Brain

Edited by John R. Shook and Tibor Solymosi

Bloomsbury Studies in American Philosophy

B L O O M S B U R Y
LONDON • NEW DELHI • NEW YORK • SYDNEY

Bloomsbury Academic
An imprint of Bloomsbury Publishing Plc

50 Bedford Square
London
WC1B 3DP
UK

1385 Broadway
New York
NY 10018
USA

www.bloomsbury.com

Bloomsbury is a registered trade mark of Bloomsbury Publishing Plc

First published 2014

British Library Cataloguing-in-Publication Data
A catalogue record for this book is available from the British Library.

ISBN: HB: 978-1-4725-1105-8
ePDF: 978-1-4725-0585-9
ePub: 978-1-4725-1368-7

Library of Congress Cataloging-in-Publication Data
A catalog record for this book is available from the Library of Congress.

Typeset by Fakenham Prepress Solutions, Fakenham, Norfolk NR21 8NN

Contents

List of Contributors

F. Thomas Burke is an American philosopher who teaches at the University of South Carolina. He is the author of *What Pragmatism Was* (2013) and *Dewey's New Logic* (1994). His work interprets contemporary philosophy of mind, philosophy of language, and philosophical logic through the lens of classical American pragmatism. He earned a BA and MA from the University of New Mexico and a PhD from Stanford University.

Eric P. Charles, PhD, is lab manager at the Center for Teaching, Research, and Learning Psychology at American University in Washington, DC. His academic work includes empirical and theoretical papers in psychology and animal behavior. Recently this has focused on resurrecting insights from Edwin Bissell Holt, who served as a bridge between William James and many of the twentieth century's most prominent psychologists. He recently published *A New Look at New Realism: The Psychology and Philosophy of E. B. Holt*, and a special issue of *The Review of General Psychology* on Unifying Theories of Psychology.

Sabrina Golonka, PhD, is a Lecturer at Leeds Metropolitan University, UK. She is a cognitive psychologist specializing in categorization and perceived similarity. Her research interests include: the effect of situational co-occurrence on similarity judgments, ecological approaches to cognition, the role of selective attention in categorization, and how the perceived similarity between people affects stereotyping and prejudice.

John Kaag, PhD, is Assistant Professor of Philosophy at the University of Massachusetts, Lowell. His work is primarily in American philosophy yet ranges across various interrelated themes. He has published several articles in the philosophy of mind, philosophy of religion, feminism, politics and technology, and neuroscience. These articles are found in noteworthy journals, including *Phenomenology and the Cognitive Sciences, Daedalus, Zygon, The Journal of Speculative Philosophy,* and *The Transactions of the Charles S. Peirce Society*. A central focus of his work is the nature of imagination and the richness of experience. His manuscript, *Thinking Through the Imagination: The Role of Aesthetic Creativity in Human Cognition*, is currently under review. He is also working on co-editing a book on pragmatism and the philosophy of sport.

Russell Pryba received his PhD in philosophy from the State University of New York at Buffalo in 2012. His main area of research is American Pragmatism, in particular the aesthetic theory of John Dewey. In addition to Dewey's aesthetics, Dr. Pryba has also written on somaesthetics and the philosophy of food. He currently teaches at Northern Arizona University.

Tim Rohrer is a philosopher and cognitive neuroscientist who has not forgotten that he is an academic with a body. One of the first researchers of his generation to develop a systematic theory of how our physical, neural, and cultural embodiment affects our linguistic and cognitive abilities, he has published work on embodiment and metaphor in fields ranging from neuroimaging to ancient philosophy, and from flight software design and robotics to political cartooning. He has held an NIH Fellowship at the Institute for Neural Computation at the Salk Institute and the University of California at San Diego, a Fulbright Fellowship to the Center for Semiotic Research at Aarhus University in Denmark, and has given invited lectures on six of the seven continents (the Antarctic penguins, for the moment, seem uninterested in embodiment theory). He is currently the Director of the Colorado Advanced Research Institute in Boulder, Colorado, and can be reached at rohrer@cogsci.ucsd.edu via email.

Maxine Sheets-Johnstone is an interdisciplinary scholar affiliated with the Department of Philosophy at the University of Oregon where she taught periodically in the 1990s and where she now holds an ongoing Courtesy Professor appointment. She began her career as a choreographer/dancer, professor of dance/dance scholar. She has published numerous articles in humanities, science, and art journals. Her book publications include *The Phenomenology of Dance; Illuminating Dance: Philosophical Explorations;* the "roots" trilogy: *The Roots of Thinking, The Roots of Power: Animate Form and Gendered Bodies,* and *The Roots of Morality; Giving the Body Its Due; The Primacy of Movement; The Corporeal Turn: An Interdisciplinary Reader*; and the recently published eBook *Putting Movement Into Your Life; A Beyond Fitness Primer.* She was awarded a Distinguished Fellowship at the Institute of Advanced Study at Durham University in the Spring of 2007 for her continuing research on xenophobia, and an Alumni Achievement Award by the School of Education, University of Wisconsin in 2011 and was honored with a Scholar's Session by the Society of Phenomenology and Existential Philosophy in 2012.

John R. Shook received his PhD in philosophy from the University at Buffalo, was a professor of philosophy at Oklahoma State University from 2000 to 2006, and then joined the faculty of the Science and the Public online EdM program at the University at Buffalo, where he also is research associate in philosophy. In recent years, he has also been a visiting fellow at the Institute for Philosophy and Public Policy at George Mason University, and at the Center for Neurotechnology Studies of the Potomac Institute for Public Policy in Virginia. Volumes he has edited include *Pragmatic Naturalism and Realism*, *the Blackwell Companion to Pragmatism*, and *The Essential William James*. His recent articles range across social neuroscience, neurophilosophy, moral psychology, and neuroethics.

Tibor Solymosi is Lecturer of Philosophy in the Department of History and Government at Bowie State University in Bowie, Maryland. He graduated *magna cum laude* from Allegheny College and earned his PhD at Southern Illinois University, where he studied at the Center for John Dewey Studies. He works at the intersections of philosophy and the sciences of life and mind. His interests have focused on the historical overlap of American pragmatism and science, and on the consequences of the sciences on our self-conception as conscious, free, and morally responsible selves. His writings focus on consciousness, free will, neurophilosophy, scientific inquiry, and ethics. They have appeared in several academic journals and interdisciplinary volumes. His website is neuropragmatism.com.

David L. Thompson was born in Ireland, received a BSc (Hons) in mathematics from the University of Alberta, and was awarded a PhD in Philosophy from Louvain University. He is professor of philosophy (retired) at Memorial University of Newfoundland. His research includes the philosophy of mind and cognitive science, phenomenology, and the philosophy of the human sciences. He has a strong interest in the philosophy of Daniel Dennett, about whom he has published a monograph and co-edited a volume of articles. His current research focuses on the normative nature of selfhood within a naturalistic, evolutionary perspective.

Mark Tschaepe, PhD, is Assistant Professor of Philosophy at Prairie View A&M University in Prairie View, Texas. He specializes in philosophy of science, ethics, and the history of philosophy.

Andrew D. Wilson is a cognitive scientist at Leeds Metropolitan University who studies the perceptual control of action and embodied cognition from an ecological approach. His primary research interests are in how we learn to use perceptual information about ourselves, and our environments to produce successful actions. His work spans simple lab-based tasks (coordinated rhythmic movement) and more complex tasks (throwing for long distances and accuracy). Dr Wilson blogs his research at http://psychsciencenotes.blogspot.co.uk/ and can be found on Twitter @PsychScientists.

1

Introduction

John R. Shook and Tibor Solymosi

I

This volume brings together some pragmatist philosophers vitally interested in neuroscience and the brain sciences generally. Pragmatism is a perennial philosophy precisely because its core views on experience, cognition, learning, knowledge, values, psychological and education development, inter-personal relationships, and social organization enjoy regular confirmation by evolutionary biology, developmental psychology, experimental sociology, and the brain sciences, including recent developments in neuroscience. Pragmatism has been "refuted" by a parade of rival philosophies (most now forgotten or feeble), but it has only been strengthened by the sciences.

Some of this volume's chapters remind us how America's first generation of pragmatists did not stumble onto its principles but instead designed them in light of novel discoveries in biology and psychology. Viewing organisms entirely naturalistically and understanding human brains in light of evolution, forced this first generation to radically re-conceive how intelligence works, during the period from the mid-1860s to the 1890s. The pragmatists were America's first cognitive scientists. Charles Peirce, William James, John Dewey, and George Mead knew the laboratory intimately. They all published their research in experimental psychology and sociology, as did many of their graduate students. They possessed few details on the specific functions of the brain's many components, but they were innocent of errors that philosophers have continued to perpetuate. For example, the pragmatists noticed how the brain grows and changes during infancy and youth, and they figured that brains grow and modify nervous connections throughout the lifetime. They presumed correctly the basic idea behind the Hebbian theory that brain tissues activated together during activity would be more likely to stay functionally related and useful for activity thereafter. They inferred that the expansion of experience and the growth of brain structure were correlated, explaining how learning consists of acquisition of workable habits for practical success in managing environing conditions. These pragmatists also thought that much of the brain's work has little to do with self-consciousness and often occurs at nonconscious levels. They therefore denied that cognition consists entirely of internal representations about static external matters rationalistically manipulated within a Cartesian theater, and they also rejected the notion that perception, cognition, emotion, and volition were discrete processes. Their psychologies were both mildly behavioristic and socially oriented, emphasizing how most distinctly human capacities from language to science depend on social training, extended cognition, experimental problem-solving, and cultural accumulation of knowledge. Outspoken on knowledge and truth, their varieties of pragmatism claimed that only social groups set working criteria for truth and produce knowledge from coordinated experimental inquiry.

Pragmatism never dominated philosophy, and it was displaced from philosophy departments by later successive empiricisms, phenomenologies, neo-Kantianisms, neo-dualisms, and analytic rationalisms. Yet pragmatism was never eclipsed where it was designed to shine and offer its transformative standpoints. Its psychological functionalism, social behaviorism, and evolutionary anthropology filtered into many of the social sciences, and permanently re-directed them during the twentieth century. Developmental psychology, experimental psychology, and cognitive science re-invigorated many pragmatist views of brain cognition, learning, and knowledge growth in the 1980s and 1990s. The first decades of the twenty-first century are accelerating this trend back to pragmatism. Since active scientists are less concerned with the history or philosophical aspects of their particular fields, it comes as no surprise that most behavioral and brain scientists are presently unaware of pragmatism. However, it is no coincidence that their theoretical advances are consistent with pragmatism, and they often find themselves defending pragmatist themes. Is this all just a coincidence, a matter of curiosity for intellectual historians?

The broad tradition of pragmatism is needed now, more than ever. As each of the chapters emphasizes in different ways, knowledge of brain function rarely stays within narrow scientific bounds, nor does it remain free from philosophical interference and commentary. Scientists should participate in those broader debates to the extent they wish, just as the classical pragmatists did, and current scientists can benefit from their earlier philosophical explorations across the same territory. But those explorations always meet staunch resistance. Rival philosophical views, some disguised within scientific paradigms and others openly espoused by armchair philosophers, offer up staunch opposition to the new pragmatist stances all over again. Defenses of core themes and theories of pragmatism based on the cognitive and brain sciences can be collectively labeled as "pragmatist neurophilosophy." When those scientifically confirmed matters are not only applied for justifying pragmatist views, but additionally formulated to dominate the field of neurophilosophy, in turn, philosophy itself, then the specialized label is Neuropragmatism.

Pragmatism in general has allied with the science-affirming philosophy of naturalism. Naturalism is perennially tested by challenges questioning its ability to accommodate and account for several philosophical questions, ranging from knowledge to personhood. Neuropragmatism is in a good position to evaluate those challenges. Neuropragmatism is no adjunct or aspect of neurophilosophy—it is its own complete neurophilosophy, and hence radically reconfigures all of philosophy. Neither pragmatist neurophilosophy nor Neuropragmatism has the unhelpful agenda of "verifying" abstract principles of pragmatism with selective neuroscience, and they do not rest their plausibility on the fads and fashions of prefixing "neuro" to any subject-matter. There is no hidden agenda for presuming that the sciences of the small (of neurons, synapses, molecules, electrons, etc.) entirely dictate what is actually happening in the world, or that the sciences can eliminate as unreal "hard" problems of perennial philosophy, such as the origin of consciousness and the normativity of reason.

Neuropragmatism does contribute to the defense of a general naturalistic outlook in philosophy against criticisms striking deeply at the coherence of any naturalistic theory of learning and reasoning. The toughest criticisms accuse naturalism of somehow denying the reality of experience and the value of consciousness, the supposed source of all information for learning, and also denying the authority of reasoning in generating knowledge. If any philosophical naturalism judges that experience is unreal or that reasoning lacks normative authority, then learning could not be possible, at least any sort of learning that naturalism would countenance.

The essays in this volume touch upon many particular aspects of these issues. Most are critical of the reductionist tendency found not only in contemporary (neuro)philosophy and science but also in popular appropriations of neuroscience. Other common themes among the chapters are a deep suspicion of Cartesianism, from the stark separation of mind/body or mind/world to the role of representation in doing cognitive science. But the commonalities do not imply that there is one single and unified platform for pragmatist neurophilosophy presented across all of the authors' contributions. Many theoretical perspectives and approaches to philosophical puzzles are offered here, informed by each author's unique background. Some authors focus on a particular thinker; others on an idea or method, or set of ideas and/or methods. Still others draw on the sciences in imaginative ways. Of course, all the contributors make use of these approaches, yet each accomplishes their results in his or her own way. Pragmatism, old and new: pragmatist neurophilosophy as a research program is quite young, but its core philosophical strategies are indebted to an old tradition with numerous intellectual contributions to its credit.

II

Part One focuses on central historical figures of pragmatism, namely Charles Sanders Peirce, William James, and John Dewey. These thinkers appear through the other chapters as well, along with pragmatic-minded thinkers such as George Herbert Mead, Richard Rorty, Hilary Putnam, and Robert Brandom. Perhaps most striking about these three initial chapters is that, although they are historical, they are also most timely, in light of advances in complexity theory, dynamic systems theory, and neuroaesthetics, which enlarge pragmatism's explanatory scope.

John Kaag highlights some of the important insights Charles Sanders Peirce made on the nature of neuronal adaptation and learning. Importantly, Kaag argues, Peirce not only anticipated Hebb's rule (that neurons that fire together, wire together), he also saw beyond Hebb's initial conception to the need for further qualifications. Kaag aligns Peirce's early insights with the mathematical views of contemporary thinkers, such as Steven Strogratz. Both thinkers endorse the view that, in order to understand how the brain works, complex and nonlinear dynamics are integral. This is a view that has largely been ignored by analytic philosophy of mind, and one that mainstream cognitive science is starting to appreciate. Kaag situates Peirce's ideas within this contemporary context such that those interested in Peirce's views on early cognitive science and those interested in what Peirce has to offer contemporary cognitive science are both unlikely to be disappointed.

Maxine Sheets-Johnstone uses William James's views about the relationship between mind and brain to combat the orthodox reductionism and brain-centrism of contemporary neuroscience, cognitive science, and philosophy. Beginning with the analogy James drew between breathing and thinking and emoting, Sheets-Johnstone guides us through James's exploration of the various intricacies and difficulties in developing a science of experience. Central to her argument is recent work in coordinating dynamics, which emphasizes both the importance of an emergentist top-down approach to causation as well as the standard reductionist bottom-up approach. That is, as James anticipated and recent research into brain-body-world dynamics is showing today, there is much more to being human and cultivating a self-understanding than what the brain or inquiry into it can tell us alone.

Finishing up Part One and setting the scene for Part Two, Russell Pryba provides an analysis of aesthetics and neuroscience from a Deweyan perspective. He begins with an elaboration on Dewey's version of naturalism, emphasizing the important relationship between science and art. With this framework in hand, he reviews the main positions in neuroaesthetics and finds them wanting. Pryba points the way forward for doing neuroaesthetics that does justice to both the lived experience of art and what neuroscience can attribute to both the experience and its further understanding.

Part Two focuses on the relationship between philosophy and neuroscience, specifically advancing the pragmatist project of reconstruction. This project is concerned with what we humans are to make of the deluge of data coming from the sciences of life and mind. The first two chapters of this section explicitly take up, develop, and apply the philosophical method of reconstruction. The spirit of these two chapters is met by the next, whose authors make a clarion call for an imaginative reconstruction of contemporary science of mind.

Tibor Solymosi provides a historical sketch of philosophical conceptions of experience, contrasting the sensationalistic empiricism of Hume (and other modern philosophers) with the radical empiricism of James and Dewey. Solymosi makes the charge that despite their efforts, both neopragmatists and neurophilosophers cannot escape the Cartesian framework behind sensationalism. This inability leads to the problem of reconciling diametrically defined projects—what Wilfrid Sellars called the scientific and manifest images. This problem of reconciliation has led to the proposed solution of neurophilosophers known as eliminative materialism. Neopragmatists reject the authority of science yet offer an eliminativism of their own: a linguistic variety Solymosi refers to as eliminative idealism. Neither neopragmatists nor neurophilosophers have made much headway in their project of reconciliation. Solymosi contends this is due to a failure to appreciate the full impact of Darwinism on the relationship between experience and the world. Solymosi contrasts the project of reconciliation with the project of reconstruction. To illustrate this contrast, Solymosi critically assesses the work of a recent popular book of neuroscience by the neuroscientist David Eagleman. Through this assessment, Solymosi argues that neuropragmatism is the way forward for pragmatism because it, like classical pragmatism but unlike neopragmatism and neurophilosophy, has the tools for critically assessing scientific data in an ameliorative fashion.

The project of reconstruction is further elaborated upon by Mark Tschaepe. In his chapter, the target is the new subfield of neuroeconomics. Tschaepe provides a useful survey of how the literature has developed on the neurotransmitter, oxytocin, from it being statistically correlated with human experiences like trust to it being the primary cause of not just trust but the entirety of human morality. Tschaepe then applies the reconstructive method to this literature. He balances the excitement of this new data with a healthy skepticism, making the case that more research can be done once we understand the role neurotransmitters play in situations having to do with humans and trust.

Empirical psychologists Eric P. Charles, Andrew D. Wilson, and Sabrina Golonka have one great hope for neuropragmatism: that it can provide a coherent language for describing the role of the brain in basic, embodied, embedded cognitive processes. An ever-increasing number of neuroscientists realize that cognitive-psychology traditions are a dead end. They tire of reading cog-neuro publications sandwiching brilliant methods and solid results between incoherent introductions and conclusions. The current mess makes it difficult to integrate research lines, and cumulative progress suffers. Charles, Wilson and Golonka argue that we neuropragmatists and empiral psychologists can offer some guidance as to the new language that is needed, for much work remains to be done. This chapter takes the reconstructive project to a more abstract level by raising questions about the very discipline of cognitive science itself, its underlying metaphors and assumptions, and subsequent aims of inquiry. Drawing on pragmatists like Peirce, James, and E. B. Holt, the authors raise a constructive challenge, not only to the whole of cognitive science but to neuropragmatism as well. Their hope is that neuropragmatists can help facilitate much needed cross-disciplinary dialog as well as critical philosophical reflection.

Part Three investigates further consequences of this pragmatist alternative to orthodox cognitive science and neuroscience. Inspired by classical pragmatism, the authors of these chapters tackle contemporary issues in cognitive science (broadly construed) and philosophy of mind. These authors point to several possible avenues of further research for neuropragmatism.

Tim Rohrer continues the constructive criticism of contemporary cognitive science by situating pragmatism and functionalism in the history and philosophy of science, particularly with regard to psychology's historical development.

This context setting also draws on the larger historico-cultural events of the twentieth century, emphasizing as Dewey often did that science takes place within a larger human context. Rohrer argues that James and Dewey's contributions to pragmatism were rooted in not just empirical science, but in their concrete observations of the human body. Rohrer continues his analysis by noting that philosophy and functionalist psychology, which once organically emerged from an embodied pragmatic standpoint, were uprooted in the latter half of the twentieth century by information theory and the conception of the mind and brain as a particular sort of digital computer. However, the cognitive sciences' recent return to the empirical methods of the cognitive neurosciences has engendered a resurgent investigation into how our embodiment shapes our thinking.

F. Thomas Burke takes an alternative route to the problem of representation to that offered by the previous two chapters. Where these authors advocate a rejection of representation because of the historical and contemporary confusions surrounding the word's use, Burke aims to reconstruct our use by appealing to mental externalism. Drawing on work in extended theory of mind, Burke bridges these contemporary arguments with the insights of John Dewey and George Herbert Mead. His reconstruction of mentation as something that goes on with our bodies in our worlds—and not something stuck in the head—offers an alternative to both orthodox cognitive science and the anti-representationalism found in varieties of pragmatism and in unorthodox positions within cognitive science.

David Thompson articulates a pragmatism that critically incorporates insights from Continental philosophy and neopragmatism, bridging Husserl and Merleau-Ponty with Brandom. To do this, Thompson begins by arguing that an organism lives in a world-for-the-organism (*umwelt*) that is defined by its needs and capacities rather than by physical principles. Even the simplest organism constitutes objects that correlate with the internal structures that unify the organism and enable perception and action. Both objects-for-the-organism and the organism itself are functional entities that are attributed their roles by the overall ecosystem. The functionality is normative and can only be understood by reference to evolutionary history. To make his case, Thompson builds a useful evolutionary heuristic from molecules to amoebas to bees and flowers. From this heuristic, he establishes the continuity between human cultures and their biology: in short, he evades the Sellarsian clash of scientific and manifest images and continues the reconstructive project of elaborating continuities. He extends his treatment of biological organisms and objects to humans, concluding that human selves live in social worlds in which objects and actions are constituted by their cultural meanings.

In the closing chapter, John R. Shook tackles a difficult question with an equally complex history. In asking whether experience is objective, subjective, or both, Shook seeks to return experience to a status amenable to scientific inquiry, as pragmatism has always expected. The way that science can investigate multiple perspectives on nature permits it to inquire into anything in human experience, including that experience, and coordinate theoretical results with each other. Scientific inquiry into experience may not look quite the way modern empiricists or Cartesian rationalists may expect. The brain sciences are discovering how and why embodied and dynamic organisms within co-responding environments generate aware experience as their nervous systems engage with outer events, and not inner neural states. Experience is therefore acquiring a respectably scientific status because the location, dimensions, and properties of experiences are amenable to detection and experimentation. Experience is all-natural, but it isn't in the head. Shook infers that, because the brain does not think about the brain, experience is not in the head, but embedded in the environing world. The smallest proper unit of experience cannot be some set of enervated brain states or any kind of representational content, but an environing situation having a live organism embedded in it. Because the sciences are able to objectively study cognition and experience, each has a valid but perspectival field of research, neither the elimination nor reduction of experience will be possible, and these sciences must coordinate their knowledge together instead.

III

Central to pragmatism is a doctrine of fallibilism. The essays collected here are among the first attempts to critically assess the impressive advances of neuroscience and their implications for both neurophilosophy and all of philosophy. Still in its infancy, the import of neuroscience cannot be taken lightly because there is much more involved here than intellectual speculation or edification. As Dewey often noted, and the authors of these chapters are keenly aware, science is a human activity performed in a human context that is not always—nor necessarily—virtuous. With President Barack Obama's calling for significant funding for the study of the brain, the need for a constructively critical perspective on neuroscience is ever greater. We believe the essays here are a very strong and impressive place to start.

The editors have several people to thank for making this volume possible. Not only do we thank the contributors, we also thank our editors at Bloomsbury, James Giordano of Georgetown University, and our ever-supportive friends and families.

Part One

Historical Considerations

2

Peirce on Neuronal Synchronicity and Spontaneous Order

John Kaag

Introduction

Analytic philosophy is, by definition, ill equipped to employ the most recent findings of contemporary neuroscience. Modes of analysis vary but hold at least one thing in common, namely the belief that subjects, arguments, and phenomena can be best understood by a study of their constitutive parts. The narrower and more refined a study can be, the greater the potential clarity concerning the broader issue at hand. This analytic approach was employed by philosophers of mind who, in the 1950s and 1960s, began to theorize on the basis of the findings of early neuroscience. The results were almost comically reductive. The study of the mind was reduced to the narrow analysis of C-fiber firings or the PET scans of particular neural localities. Today, most scientific researchers accept that brain processes, along with most other biological systems, cannot be accurately described as a sum of their constituents. That is to say, we know that biological systems are not best understood by way of analysis.

The studies of neural complexity, multi-modality, and degeneracy indicate that the spontaneous order exhibited in brain processes is an emergent phenomenon that cannot be described as a simple summation of parts. This finding presents an existential either-or to analytic philosophy of mind: either philosophers of mind can amend their commitment to the method of analysis, or they can willfully ignore the most forward-looking aspects of contemporary neuroscience. To this point, with only a few exceptions, philosophy has opted for willing ignorance. This chapter argues that another strain of philosophy, largely overlooked in the past century, is better equipped to learn from and inform the studies of neural complexity and emergence. American pragmatism, initiated by C. S. Peirce, and developed by William James and John Dewey, never took analysis as the final resting place of philosophical reflection. Instead, they insisted that phenomena should be synthetically (as well as analytically) understood. This insistence stems, at least in part, from their belief that philosophy, at its best, is to work at the reconstruction of experience, a phenomenon that refuses to be easily reduced. To say that experience cannot be analytically grasped, however, is not to eschew theoretical rigor but rather to demand that our precise explanatory models account for the global and holistic character of the phenomena in question. Peirce was especially adamant on this point, anticipating many of this century's general system theorists in their claims that the most sophisticated mathematical techniques would have to be employed to "get at" the complexity of phenomena.

Listening to a bluegrass band, playing basketball with an old friend, rowing on the Charles River, teaching a crew of undergraduates, feeling the purr of a well-tuned engine: this is the stuff of life. More specifically, this is the stuff of human life. The pulsating give-and-take of experience is the place where human meaning, at its most rudimentary and its most complex, happens. What is it about the pulse of experience—the repetition that keeps things on the move? I was once told that the term experience could not do the heavy philosophical lifting that thinkers such as Dewey and James wished it to do. I disagree. I believe that the complex phenomenom of experience invites us to find equally complex modes of investigation in order to understand it. Attending to the multi-modal flow of experience not only allows us to explore the existential ground of human life but also encourages philosophers to delve into new fields of science that have begun to re-examine the biological and physical preconditions of meaningful human activity.

This focus on experience, and on the particular ways that experience "plays out," was at the heart of Charles Peirce's philosophical project. Peirce comments that "experience is our only teacher" in the sense that only through experience can we learn things by heart, but also in the sense that the complexity of human experience ought to remain the topic of scientific and theoretical investigation.[1] One of the characteristics of human experience is, in the words of Steven Strogatz, its "synch." Strogatz, an applied mathematician and complexity theorist at Cornell, writes that, "We take for granted that we can sing together and dance together, march in step, clap in unison."[2] The ability to spontaneously synchronize with one and other, the tendency to move, speak, smile, row, play, speak together, etc. is second nature to us. "But because it comes so easily we have poor insight about what it actually demands."[3] What occurs at the moment when things get into synch? What are the biological preconditions of human synch? We rightly insist that human synch requires at least a low level of human intelligence, the ability to tie our behavior and anticipate that of others. Examples of non-human living synchronicity (the chorusing of crickets for example) are often attributed to very rudimentary forms of intelligence and communication. More often than not, according to Strogatz, we chalk the marvels of living synchronicity up "to the marvels of evolution, the magic of millions of years of natural selection."[4] But here we fail to ask what sort of awesome complexity and emergence underlies our present moment. In so doing, we run the risk of deferring the problem of synchronicity, or at least ignoring its real mystery, namely what happens at the edge of chaos and spontaneous order. That is the question that this chapter will take up, focusing on the way that Peirce's philosophy is amenable to, and in many cases anticipates, the findings of complexity theorists.

Experience, habit formation, and Hebb's rule[5]

How does experience happen? There is a phenomenological answer to this question that tries, as best it can, to get to the qualitative dimension of human experience. There is also, obviously, an empirical answer that has traditionally missed the feeling of experiential qualia. And I take Colin McGinn's argument concerning the reason why cognitive neuroscience will never get a handle on the quality of an experience.[6] This being said, if we identify experience with the phenomena of synch, and then ask what are the general characters of synch at a variety of levels of empirical analysis (in this case at the level of neural coordination) we might get a better sense of the biological origins of experience and the nested embodied character of mind.

According to American pragmatists like Peirce, James, and Dewey, it does not happen as a series of random inputs as Humean or Lockean empiricists suggested or as the clear and distinct ideas of the Cartesian rationalists. Following Kant, the pragmatists attempted to mediate between the two dominant schools of modern philosophy, and along these lines, they maintained that experience happened as the reciprocal transaction between the external world and the human subject. This reciprocity meant that the world has the chance to affect our embodied minds, but that our bodily comportment also structures and receives the world in particular ways. This is what Dewey meant when he suggested that experience is a process of undergoing and doing.[7] There is always a "human contribution" in experience, but there is also, and equally, a contribution that the world makes in order to round out experience. This rough sketch of experience fits nicely with phenomenological descriptions of perception and cognition that hold there is no radical divide between mind and body or mind and world. The question that Peirce asked, and that anticipated the cognitive neuroscientists of the last 50 years, was the following: what are the anatomical preconditions for human organisms to receive the world of experience?

Peirce believed that any adequate answer would have to address two crucial aspects of human cognition. First, it would have to account for the robust role that habit plays in our negotiation of experience. Second, it would have to outline the conditions under which habit, uniform reactions to a given stimulus, might prove plastic enough to respond in adaptive ways to new stimuli. From the point of view of the complexity theorist, such as Strogatz (and Peirce), these are very hard questions to answer in a rigorous way. The second issue, that of plasticity and adaptation (how habits are revised and overcome), is particularly difficult, for the models that best describe it are nonlinear systems that reflect emergent properties. Before turning to this difficulty, a preparatory, and much easier question can be asked. At the level of human observation, how are habits formed in the embodied mind of a human organism? In 1880, ten years before his friend William James gave a more detailed rendering of habit in *The Principles of Psychology* (1890), Peirce describes habit as a function of synaptic activation. He does so in a long passage that deserves careful attention:

> Thinking, as cerebration, is no doubt subject to the general laws of nervous action. When a group of nerves are stimulated, the ganglions with which the group is most intimately connected on the whole are thrown into an active state, which in turn usually occasions movements of the body. The stimulation continuing, the irritation spreads from ganglion to ganglion (usually increasing meantime). Soon, too, the parts first excited begin to show fatigue; and thus for a double reason the bodily activity is of a changing kind. When the stimulus is withdrawn, the excitement quickly subsides … It results from these facts that when a nerve is affected, the reflex action, if it is not at first of the sort to remove the irritation, will change its character again and again until the irritation is removed; and then the action will cease. Now, all vital processes tend to become easier on repetition. Along whatever path a nervous discharge has once taken place, in that path a new discharge is the more likely to take place. Accordingly, when an irritation of the nerves is repeated, all the various actions which have taken place on previous similar occasions are the more likely to take place now, and those are most likely to take place which have most frequently taken place on those previous occasions.[8]

This passage is an early version of what would come to be called Hebb's rule, a key insight of modern cognitive neuroscience. In 1949, Donald Hebb wrote *The Organization of Behavior* in which he postulates: "When an axon of cell A is near enough to excite a cell B and repeatedly or persistently takes part in firing it, some growth process or metabolic change takes place in one or both cells such that A's efficiency, as one of the cells firing B, is increased."[9] In short, when neurons wire together, they fire together. According to Hebb, this correlation can be mathematically described in the following manner:

$$W_{ij}=X_iX_j$$

where W_{ij} is the weight of the connection from neuron j to i and X_i the input for neuron i.

This equation can be re-written in the following way that will allow us to analyze its strengths and shortcomings:

$$\tau_w \frac{d\mathbf{w}}{dt} = v\mathbf{u},\ \mathbf{w}_{t+1} = \mathbf{w}_t + \frac{\delta t}{\tau_w} v\mathbf{u}$$

With the development of this function, Hebb lays the groundwork for connectionist theorists in the 1960s who began to create models that envision brain functions as emergent with global properties that result from the interaction of connected neural networks. Hebb's rule is also significant in the sense that it identifies growth and metabolic change as the defining features of our nervous system, a feature that coincides with the spontaneity and creativity that has been associated with human experience. Several points, however, need to be highlighted in order to recognize and qualify the import of Hebb's insight.

First, the metabolic growth (ΔT_{AB}) is calculated as a function of the instantaneous interaction of two individual cells. It is the case that growth depends on the actions and reactions of *both* cells and emerges as a process that is irreducible to the excitation of either A or B in isolation. What some interpretations of Hebb's rule neglect, however, in their attempt to remain parsimonious, is the fact that these two cells are *already* interacting in wider neuronal complexes, which set the limits and provide the enabling conditions of the cells' excitation. This limit and enablement arises in the ongoing and non-repeatable transformation of brain states and metabolic processes. Gerald Edelman points to these processes in his description of the creation of secondary neural repertoires.[10] In emphasizing this point, we can see that Hebb's rule cannot be easily applied to the activation dynamics of neural populations, since these complex dynamics cannot be modeled by a mechanical rule or linear function. A second point needs to be emphasized in regard to Hebb's rule: the metabolic growth rate is calculated as change at a particular instant in time. This calculation provides only a snapshot or glimpse of neural dynamics without demonstrating the course of events and relations that might have led to this rate of change. That is also to say that the Hebb's rule provides an idealized model of an extremely small time interval of continuous and diverse evolutionary, developmental, and experiential processes. This is not to dismiss Hebb's work, but to encourage us to extend and modify the rule in order to better demonstrate the continuity of neural activation and development. This qualification of Hebb's rule suggests that neural dynamics should not be regarded as mechanical linear functions. This said, the rule also does not support the idea that these dynamics are the products of random chance or pure contingency. To return to a point made earlier, *this form of creative irregularity is irreversible and directed.* As psychiatrist Jeffrey Schwartz notes in his experiments on language acquisition, "Once the Hebbian process has claimed circuits, they are hard-wired for that sound; so far, neuroscientists have not observed any situations in which the Hebbian process is reversed so that someone born into a sea of, say, Finnish loses the ability to hear Finnish unique sounds."[11] The metabolic change that occurs is an irrevocable fact that continues to affect—albeit in a dissipating or decaying degree—the future transformations in the neural activation of the system. It is in this sense that neural dynamics are motivated by, and arise in, the creative spontaneity that has been outlined in our discussion of the imagination, and in Peirce's description of abduction and tychism.

Hebb's hypothesis, if modified with these qualifications, is born out in modern cognitive science and is consistent with Peirce's understanding of habit formation, but it also points to the way in which novel input stimuli can revise these habits of association. This formula, however, is misleading in its simplicity. In the years after Hebb's discovery, scholars distilled two problems with Hebb's theory from the above discussion. First, Hebb describes no process by which connections can get weaker over time. Second, Hebb's rule provides no upper bound or limit for how strong connections can get. This is to say that Hebb's rule was unstable. It is obvious from the long passage given above that Peirce was very concerned about this instability. The first half of the passage deals with the depression of synaptic activation, and the second half of the passage deals explicitly with the increase of synaptic weight. And this is where what appears easy at first gets very, very hard. Bienenstock, Cooper and Munro (1982) proposed a model (BCM) of synaptic modification that holds that neurons possess a synaptic modification threshold that is not fixed but instead varies according to a nonlinear function with the average output of the cell.[12]

$$\tau_w \frac{d\mathbf{w}}{dt} = v(v-\theta_v)\mathbf{u}$$

$$\tau_{\theta v} \frac{d\theta_v}{dt} = v^2-\theta_v$$

$$\tau_w < \tau_{\theta v}$$

More recently, Byrne translated this model into a variety of quantitative expressions in which a neuron A with average firing weight V_A, projects to neuron B with average firing weight V_B. The synaptic connection from A to B has strength T_{AB}, which determines the degree of activity in A is capable of exciting in B; the strength of T_{AB} should be modified in some way that is dependent on *both* activity of A and B. The general expression of this plasticity rule is formulated in the function:[13]

$$\Delta T_{AB} = F(V_A, V_B)$$

How is one to model this function, or more specifically, the activation patterns whereby a conglomerate of neurons coordinate and channel over a period of time? One answer will take us deep into the field of complexity in its study of nonlinear dynamics. It is a field that Peirce stumbled into and explored with surprising accuracy well before thinkers like Strogatz took over. Long before Bienenstock, Cooper and Munro revised Hebb's rule, Peirce was on the trail of a model of neural synchronization and found a mathematical model in his work on pendulum studies in the 1880s.

Coupled oscillation in simple and complex systems

In the last decade, Strogatz has dedicated himself to studying the phenomena of synchronization of living and non-living systems, a dynamic whereby agents coordinate their movements over a period of time. When Hebb finds that "neurons that fire together, wire together" he is, in a very elementary way, gesturing to two vital aspects of neuronal synchronicity. First, he is pointing to the fact that neurons and neuronal groups function as oscillators whose frequency varies depending on the frequencies of their neuronal neighbors. Second, he is suggesting that these neuronal groups function as a network that can "save" information depending on the channeling of particular frequencies in a population. The first aspect of this synchronicity in non-living actors was described by Christiaan Huygens in 1665 when we observed that two pendulum clocks, when placed on the same wall will eventually fall into time, swinging at precisely the opposite phase. He hypothesized—and it was later proven—that this phenomenon could be attributed to the imperceptible motion of the wall beam itself, caused by the motion of the two pendulums. This imperceptible force caused global synchronicity to the pendulum system. As Peirce began his employment with the Geodetic Survey in the 1870s, precisely at the time when he was thinking through the comment made in thinking as cerebration that anticipates Hebb, he began to examine the phenomenon that Huygens had discovered. He was asked to investigate the source of error in pendulum studies and found that it could be traced to the way that oscillators can affect each other's activities in a bidirectional manner. It was not until 1893, however, in the "Connection Between Matter and Mind," that he wrote the following long passage concerning these experiments and their implications on cognitive science. He writes:

> A clock with a heavy pendulum had been solidly fastened to a brick wall. About six feet away was a shelf attached to the same wall. A Hardy's noddy, or little inverted pendulum swinging by the bending of a spring, having been adjusted to the period of oscillation of the clock-pendulum was placed upon that shelf, when it immediately began to swing in unmistakable synchronism with the clock; and this motion it kept up for months. Again, two pendulums exactly alike and weighing some 20 pounds each, were suspended from a two-inch plank which was supported all along its eighteen-inch ends by quarter-inch plates of vertically corrugated iron attached to the floor. The direction of swinging of both pendulums was that of the length of the plank. Both being at rest, one of them was started into oscillation. Almost immediately, the other began to swing, and at the end of about three minutes the one first started had come to rest, while the one that had not been touched was swinging with nearly the same amplitude that the first had in the beginning. But no sooner had the latter come to rest than it began to start up again, while the arcs of the other became less; and in three minutes more, substantially the original condition of things had been restored. Of course, the real explanation of these phenomena is that the support of the heavy pendulum the brick wall in the one case, the corrugated iron in the other—yielded under the swaying of the pendulums, and so communicated an impulse from each to the other, increasing the motion of the one that was behind and diminishing that of the one that was ahead; and though this effect was in itself imperceptible, yet when it had been multiplied

one or two hundredfold, or as many times as the repetitions of its cause, the total effect became great.[14]

In this case, Peirce is investigating a very simple synchronous system of two actors. Even with such a simple model, we already need to use differential equations to describe the dynamic, those equations that proved so helpful in describing rate of change. Because we now have two oscillators, we will have two phases (θ_1 and θ_2) and two natural frequencies (ω_1 and ω_2) to account for. If we assume that the natural frequencies are fixed, then we will need two equations for the two unknowns θ_1 and θ_2. The pendulum has phase θ_1 and frequency ω_1. The second has phase θ_2 and frequency ω_2. For both pendulums to swing in sync with one another, the two thetas must be equal to one another. Mathematically, $\theta_1 - \theta_2$ must equal zero. The phase difference, $\theta_1 - \theta_2$, determines the extent of "correction" each pendulum needs to make to synchronize with the other one. The necessary adjustment varies depending on how far apart the two pendulums are in their cycles. If the two pendulums are very far apart in their cycles, a large correction is needed. If they are only slightly out of sync, only a slight adjustment is required. However, the situation is a bit more complex than this. The difference in θ is what influences the pendulums as to what to do. A function capable of modeling either a speed up or a slow down must be able to periodically take on positive or negative values, depending on the difference in θ. Once again, this is ideally a sinusoid. So our mathematical model of how a pendulum adjusts its cycle to achieve synchronization with another should look something like this:

First pendulum: $\frac{d\theta_1}{dt} \approx \sin(\theta_2 - \theta_1)$

Second pendulum: $\frac{d\theta_2}{dt} \approx \sin(\theta_2 - \theta_1)$

The sine function should be mediated by a constant (K) that represents the strength of interaction between the two pendulums. We can designate this constant K_1 for the first pendulum and K_2 for the second pendulum. Additionally, we need to factor in the natural or initial frequency of the pendulums (ω_1 and ω_2 respectively), yielding:

First pendulum: $\frac{d\theta_1}{dt} = \omega_1 + k_1 \sin(\theta_2 - \theta_1)$

Second pendulum: $\frac{d\theta_2}{dt} = \omega_2 + k_2 \sin(\theta_2 - \theta_1)$

Now, in the case of synchronicity, $\theta_1 - \theta_2 = 0$, and we can let $\varphi = \theta_1 - \theta_2$ to introduce a single, convenient variable to represent the phase difference. The change in φ, representing how the phase difference changes, would then be:

$$\frac{d\phi}{dt} = \frac{d\theta_1}{dt} - \frac{d\theta_2}{dt}$$

Or written, using the equations above, this would be expressed in the following way:

$$\frac{d\phi}{dt} = \omega_1 - \omega_2 - (k_1 + k_2 \sin\phi)$$

All of this was mathematical child's play for Peirce, who explains that what appears to be an odd sympathy between the two pendulums can be understood through the calculus of continuity (i.e. differential calculus). He completes his 1893 analysis of the two pendulums in the following manner:

> Meantime (the pendulum-support in such experiments, not being seen to sway, and being recognized to be almost quite rigid) each pendulum always looks as if it were going through its curious succession of motions of its own accord, although in a sort of inverse sympathy with the other. Nor is this a mere ocular illusion; for when the differential equations which define the forces are integrated, it is found that the motion of each pendulum is simply the sum of two perfectly simple and regular funipendulous motions of slightly different periods, and is, in fact, exactly like that of a cork floating upon still water traversed by two series of waves of slightly different lengths. The remarkable thing is that the motion of each pendulum is expressed as perfectly regular, without reference to the other. And what is so manifest in regard to pendulums is equally true of any other dynamical phenomenon. Namely, the forces are expressed by differential equations; but those equations have their integrals, whether mathematicians have yet discovered them or not; and these integrals express the motion of each particle as perfectly regular and determinate, and express it independently of all other particles.[15]

It is clear that Peirce has a particular conclusion in mind when he finishes his discussion of these two pendulums. He is suggesting that while matter and mind appear to be orienting themselves independently, in a type of psychophysical parallelism, they are actually coordinating with each other, synchronizing their rates of change continuously. He makes this point again in his criticism of C. A. Strong's *Why the Mind has a Body* (1903). I do not, however, think that this is the only or most promising way that Peirce understood his studies of synchronization, as evidenced by his repeated remarks on the concept of habit, remarks that resonate so closely with Hebb's rule. What Peirce is brushing up against is the fact that neurons are oscillators and perform in similar ways to pendulums, but that they coordinate, not in isolated pairs, but in massively complex systems that can only be described by nonlinear differential equations. He seems to realize something that Strogatz articulates quite nicely, namely that "the hitch though is, that linear systems (like the two pendulums) are incapable of rich behavior. The spread of infectious disease, the intense coherence of a laser beam, the roiling motion of a turbulent fluid: all of these are governed by nonlinear equations." We would add neuronal group dynamics to this list of nonlinear occurrences since, as Strogatz asserts, "Whenever the whole is different from the sum of the parts—wherever there's cooperation or competition going on—the governing equations must be nonlinear."[16] Peirce was intent on applying differential calculus to these nonlinear systems, noting that his intent in his later writings was to articulate the "idea of continuity, or unbrokenness" in systems of ever greater complexity. Peirce suggests the tight relationship between this concept and the mathematics that seized his attention since his days at the Survey: continuity "is the leading idea of the differential calculus and of all the useful branches of mathematics; it plays a great part in all scientific thought, and the greater the more scientific that thought is; and it is the master key which adepts tell us unlocks the arcana of philosophy." Peirce realized that oscillators such as the pendulums of the simple system described earlier can best be described as coupled differential equations, and his published and unpublished manuscripts suggest that he was intent on a complex system of oscillators which, as researchers have observed in the last 20 years, reflects a tendency toward self-organization. Indeed, Peirce was following his father's lead in pursuing this belief since Benjamin Peirce had published a 600-page text entitled *A System of Analytic Mechanics* (1872), which detailed the findings of Hamilton and Lagrange, pioneers in nonlinear dynamics.[17] In the decades following his father's analysis, Peirce would remain deeply interested in models that described the spontaneous ordering of natural systems. His ordering occurred, according to Peirce, in the following manner:

> The hypothesis suggested by the present writer is that all laws are results of evolution; that underlying all other laws is the only tendency which can grow by its own virtue, the tendency of all things to take habits. Now since this same tendency is the one sole fundamental law of mind, it follows that the physical evolution works toward ends in the same way that mental action works toward ends, and thus in one aspect of the matter it would be perfectly true to say that

> final causation is alone primary. Yet, on the other hand, the law of habit is a simple formal law, a law of efficient causation; so that either way of regarding the matter is equally true, although the former is more fully intelligent. Meantime, if law is a result of evolution, which is a process lasting through all time, it follows that no law is absolute. That is, we must suppose that the phenomena themselves involve departures from law analogous to errors of observation.[18]

In this passage, Peirce is struggling, once again, with the thorny issue of determinism. We should remember that, in Peirce's treatment of the pendulums, he wrote that the integrals of the differential equations "express the motion of each particle as perfectly regular and determinate, and express it independently of all other particles." In his later writings, Peirce is eager to explore the extent to which this hypothesis can be applied to complex nonlinear systems, the sort of systems that might model neuronal activity. He suggests that these complex systems involve phenomena that depart "from law analogous to errors of observation."[19] This is not to suggest that departures from law or regularity merely seem like departures to us (due to error) but rather that no law can be provided to exhaustively define their activity. That being said, dynamical systems frequently channel or conform to patterned behavior, that is to say, move from chaos to order. And this process can be explained in terms of coupled oscillation in a population of actors, what Strogatz calls "synch." This is the process of habit formation that Peirce spent so much of his philosophical career worrying about. Connecting Peirce's comments concerning habit formation and his work on dynamical systems in his scientific experiments will, I believe, enrich our understanding of Peirce and help neuroscience avoid the reductionism that it often tends toward. Ordered behavior in natural systems according to Peirce was neither the function of some sort of deterministic behavior nor the sort of free will or agency as they are traditionally understood (a single autonomous actor carrying out purposeful tasks over time). Instead synch can occur through a type of agency that is a function of the interplay of chance and selection:

> But the writer has not supposed that this phenomenon had any connection with free will. In so far as evolution follows a law, the law of habit, instead of being a movement from homogeneity to heterogeneity, is growth from difformity to uniformity. But the chance divergences from law are perpetually acting to increase the variety of the world, and are checked by a sort of natural selection and otherwise (for the writer does not think the selective principle sufficient), so that the general result may be described as "organized heterogeneity," or, better, rationalized variety. In view of the principle of continuity, the supreme guide in framing philosophical hypotheses, we must, under this theory, regard matter as

> mind whose habits have become fixed so as to lose the powers of forming them and losing them, while mind is to be regarded as a chemical genus of extreme complexity and instability. It has acquired in a remarkable degree a habit of taking and laying aside habits. The fundamental divergences from law must here be most extraordinarily high, although probably very far indeed from attaining any directly observable magnitude. But their effect is to cause the laws of mind to be themselves of so fluid a character as to simulate divergences from law.[20]

In order to get a better sense of how this "organized heterogeneity" takes place, we can turn our attention to Strogatz's recent work on "synch." Strogatz directs his reader back to the work of Norbert Wiener whose *Cybernetics* (1950) revolutionized the theory of random processes. Following Wiener, Strogatz asks his reader to imagine a population of oscillators (neurons) that oscillate (fire) at random frequencies. Assume that this population of oscillators are normally distributed in terms of frequency around a mean of 10, in a range from 6 to 14. If these neurons are connected, however, meaning that the frequency of one neuron affects the frequency of its neighbors, the population will eventually "spike" around the mean frequency (the question will be how "connected" these oscillators have to be in order to synch). The random firing of the population will coalesce and order itself around a central attractor. This occurs since the greatest number of neurons are already firing close to the mean and attract more distant deviations toward this common frequency. Only the outliers of the neuronal population are distant enough to resist this attraction. Andrew Reynolds and others have noted that this model of convergence was anticipated by Peirce in his conception of the Real or the movement toward "truth in the long run."[21] I am not making this point. Rather, I am suggesting that the biological processes of neuronal populations underpin this unique dynamic of inquiry, that our bodies and nervous systems are structured in such a way that this convergence of opinion is possible in the manner that Peirce described.

Before closing, a brief qualification needs to be made. It is not the case that every population of oscillators will synchronize or even partially synchronize. If the populations are large enough and the frequencies of the oscillators varied enough, the activity of the population as a whole will remain "incoherent," in other words, in a state in which all oscillators drift randomly independently from one another. Large parts of Strogatz's career were spent working through the mathematical models that can describe systems that are either incoherent, partially synchronized, or fully synchronized. One of the most elegant models was developed by Yoshiki Kuramoto, who makes a few simplifying assumptions about the population of actors. First, he assumes that all oscillators are linked to all other oscillators; that these oscillators are nearly identical; and that the interaction between the oscillators depends sinusoidally on the phase difference between the two oscillators. With these assumptions, he developed the following equation where omega is the natural frequency of each oscillator and K is the coupling constant. This nonlinear equation can be solved for all natural frequencies (a Lorentzian distribution, for example) and in all cases, Kuramoto discovered that there was a K_c above which synchronization can occur and below which synchronization cannot occur.

$$\dot{\theta}_i = \omega_i + \frac{K}{N}\sum_{j=1}^{N} \sin(\theta_j - \theta_i),\ i = 1, \ldots, N$$

In 2007, Cummin and Unsworth generalized the Kuramoto model to analyze neuronal activation. They argued that since the neurons of the brain are connected in a vast number of ways that the Kuramoto could be applied with minor adjustments to its original formulation. The Kuramoto model's coupling strength (K) is time-independent, so slight alterations had to be made to accommodate the time-dependent coupling strength of neuronal activity.[22]

One of the most philosophically significant aspects of this mathematical research is the way that it attempts to model the limits between chaos and order, delimiting a space that cannot be described fully by either of these terms in isolation. What is truly interesting about nonlinear systems is not that they sometimes channel, attract, or aggregate in order to become stable, but *how* and *under what conditions* they do so. This is a shift in emphasis that contemporary cognitive scientists are beginning to take note of. Peirce was not primarily interested in the fact that habits dominate our world of experience (although he occasionally seems to be), but rather he was fascinated by the process by which certain habits come to have sway in our bodies and minds, and by the same token, he was fascinated by the sudden, often unexplainable, ways that habits are revised and overcome. The mathematical models above and the research in synchronicity gesture toward new ways to articulate this very old philosophical interest.

Conclusion

Recent work on nonlinear dynamics and synch extends Peirce's intuitions concerning the law of mind in important ways, and together they can help contemporary neuroscientists reorient the questions that they ask about neural structures and processes. What is most significant about reading Peirce in tandem with contemporary complexity theorists is the way that their writings suggest that attempts to understand neuronal activity in terms of discrete and separate processes (like isolating particular neural correlates for behavior) only captures a small fraction of the complex and nonlinear dynamics that continuously underlie human experience. Indeed, such analytic treatments of neuroscience risk distracting us from the larger questions of how best to describe genuine continuity (to use Peircean language) or genuine complexity (to use more modern verbiage). Neuronal activity is unequivocally complex. Complex systems can and do act in such a way that they seem to be unified and synchronous—a single body or community acting as one. But this unity belies a deeper and more meaningful reality, namely that an array of actors came into synch under a variety of conditions in that fertile space between chance and order. This was a space where Pierce suggests that thoughtful life happens and that we overlook it at our own peril.

Notes

1 Peirce 1893–1913, p. 153.
2 Strogatz 2012, p. 108.
3 *Ibid.*
4 *Ibid.*
5 Parts of this section will also appear in Kaag 2013, p. 147.
6 McGinn 1989, pp. 349–66.
7 Dewey 2005, p. 49.
8 Peirce 1867–93, p. 201.
9 Hebb 1949, p. 45.
10 Edelman 2005, p. 201.
11 Schwartz and Begley 2002, p. 118.
12 Elie Bienenstock Leon N. Cooper, and Paul W. Munro, et al. 1982, pp. 32–48.
13 Byrne 1989.
14 Charles Peirce, *The Collected Papers of Charles Sanders Peirce*, vols 1–6, 1931–5, Charles Hartshorne and Paul Weiss (eds); vols. 7–8, 1958, ed. Arthur W. Burks (Cambridge, MA: Harvard University Press): CP 6:273. Henceforth, CP, followed by volume number and paragraph number.
15 CP 5:179.
16 Strogatz 2012, p. 51.
17 Peirce 1872.

18 CP 6:101.
19 *Ibid.*
20 *Ibid.*
21 Reynolds 2000, pp. 287–314.
22 This section on Kuramoto is a distillation of one section of Strogatz's analysis in Steven Strogatz, "From Kuramoto to Crawford: exploring the onset of synchronization in populations of coupled oscillators," *Physica D* 43 (2000), pp. 1–20, specifically pp. 3–4.

References

Bienenstock, E., Cooper, L., and Munro, P. (1982), "Theory for the Development of Neuronal Selectivity: Orientation specificity and binocular interaction in the visual cortex," *The Journal of Neuroscience*, 2(1): 32–48.

Byrne, J. (1989), *Neural Models of Plasticity*. San Diego, CA: Academic Press.

Dewey, J. (2005), *Art as Experience*. New York: Penguin Books.

Edelman, G. (2005), *Neural Darwinism*. New York: Basic Books, p. 201.

Hebb, D. (1949), *The Organization of Behavior: A Neuropsychological Theory*. New York: Wiley.

Kaag, J. (2013), *Thinking Through the Imagination*. New York: Fordham University Press.

McGinn, C. (1989), "Can We Solve the Mind Body Problem?" *Mind*, 98: 349–66.

Peirce, B. (1872), *A System of Analytic Mechanics*. New York: D. Van Norstrand.

Peirce, C. (1931–5), *The Collected Papers of Charles Sanders Peirce*, vols 1–6, Charles Hartshorne and Paul Weiss (eds); vols 7–8, 1958, ed. Arthur W. Burks. Cambridge, MA: Harvard University Press, CP 6: 273.

Peirce, C. S. (1867–93), "On the Algebra of Logic," in *The Essential Peirce: Selected Philosophical Writings, Volume 1*, Nathan Houser and Christian Kloesel (eds). Bloomington and Indianapolis, IN: Indiana University Press.

—(1893–1913),"On Phenomenology," in *The Essential Peirce: Selected Philosophical Writings, Volume 2*, ed. The Peirce Edition Project. Bloomington and Indianapolis, IN: Indiana University Press.

Reynolds, A. (2000), "Statistical Method and the Peircean Account of Truth," *Canadian Journal of Philosophy*, 30(2): 287–314.

Schwartz, J., and Begley, S. (2002), *The Mind and the Brain: Neuroplasticity and the Power of Mental Force*. New York: Regent Publishing.

Strogatz, S. (2000), "From Kuramoto to Crawford: exploring the onset of synchronization in populations of coupled oscillators," *Physica D*, 43: 1–20.

—(2012), *Synch: How Order Emerges from Chaos in the Universe, in Nature, and Daily Life*. New York: Hyperion.

3

The Legacy of William James

Lessons for Today's Twenty-first Century Neuroscience

Maxine Sheets-Johnstone

[N]ature in her unfathomable designs has mixed us of clay and flame, of brain and mind, ... the two things hang indubitably together and determine each other's being, but how or why, no mortal may ever know.

William James[1]

And if a brain could be self-conscious, it would be conscious of itself as a brain and not as something of an altogether different kind.

William James[2]

Introduction

Twenty-first-century neuroscience is in thrall to *the brain*—the putatively incomparable human one. Its thralldom is an impediment to both evolutionary and existential understandings of life, understandings that William James spelled out in reflective breadth, in empirical detail, and with exceptional precision. In particular, James addressed questions concerning consciousness, thinking, voluntary action, the brain, emotion, self, attention, and much, much more. He did so in a rigorous way, with consummate recognition and discussion of conflicting theories and beliefs. His findings justify a substantive and penetrating critical assessment of today's neuroscience that, with its reductionist stance and experiential ascriptions to *the brain*, ignores both corporeal matters of fact—such as the fact that the brain is part of a whole-body nervous system—and basic evolutionary realities—such as the fact that, as James observed, "All nervous centres have ... in the first instance one essential function, that of 'intelligent' action."[3]

While notable discoveries have been made in neuroscience, many if not most present-day brain scientists are tethered to a reductionist ideology and thereby fail to recognize the wider but basic truths that James's extensive research uncovered. These truths have actually been vindicated by non-ideologically tethered brain scientists such as Roger Sperry—who concluded that the brain is an organ of and for movement—and J. A. Scott Kelso who, through his studies of coordination dynamics, draws attention in explicit ways to the complementarity of mind and brain. The present chapter shows how just such findings and studies support James's fundamental insights into thinking as an organic phenomenon and into feeling as both an emotional and kinetic phenomenon, and in addition how they vindicate his correlational understanding of mind and brain. In drawing on such researchers, the chapter in the end shows how the legacy of William James contains lessons from which present-day neuroscience might well prosper.

Breath: The background neurophysiology

While commonly described as simply a function or functional phenomenon endemic to living beings, the existential reality of breathing resonated far more deeply for William James. "Let the case be what it may in others, I am as confident as I am of anything that, in myself, the stream of thinking (which I recognize emphatically as a phenomenon) is only a careless name for what, when scrutinized, reveals itself to consist chiefly of the stream of my breathing." James went on to describe the relationship as follows:

> The "I think" which Kant said must be able to accompany all my objects, is the "I breathe" which actually does accompany them. There are other internal facts besides breathing (intracephalic muscular adjustments, etc. [...]) and these increase the assets of "consciousness," so far as the latter is subject to immediate perception; but breath, which was ever the original "spirit," moving outwards, between the glottis and the nostrils is, I am persuaded, the essence out of which philosophers have constructed the entity known to them as consciousness. *That entity is fictitious, while thoughts in the concrete are fully real. But thoughts in the concrete are made of the same stuff as things are.*[4]

The stream of breathing and the stream of thinking in fact share some remarkable commonalities. James did not specify these, but they warrant attention—and reflection. To begin with, as the physiological website Control of Breath notes at the very start of its article, "Although the respiratory muscles can be controlled voluntarily, normal breathing is a rhythmic, involuntary act that continues even when a person is unconscious." Like breathing, thinking too is involuntary as well as voluntary. Indeed, thoughts simply arise, not only as when an idea "pops into our head" but as when seeing something we might like, we spontaneously begin thinking about the possibility of obtaining it or buying it, or when something new arises—an invitation to a party or the possibility of a promotion at work, for example—thoughts spontaneously roll forth about it. In short, while we can and certainly do apply ourselves to thinking—thinking through itineraries in planning a trip, thinking through a budget problem in trying to cut expenses—we are at bottom, as it were, at the mercy of our thoughts. We can indeed control them, but they nevertheless retain a certain autonomy. They are from this perspective not only very much like breathing, but like emotions. Emotions too simply arise. As Jung cogently remarks, emotion "is not an activity of the individual but something that happens to him."[5] Moreover just as we can both restrain and feign emotions, we can similarly restrain and feign thoughts, restrain ourselves from thinking certain thoughts, though it must be added that the thoughts can tug at their leash, so to speak, and continue to bark at the door. Similarly, we can feign thoughts as when we attempt to convince others and even ourselves that we hold a certain view or opinion.

Inspiration and expiration are a further commonality, and not just by linguistic coincidence. Inspirations bring something new in the way of fresh ideas as well as fresh air. A new idea that comes to mind coincides in fact with an intake of breath. Expirations empty, let go of something inside, as in a sigh of relief—or laughter—as well as in stale air. When we find something not the case that we feared might be the case, we let our breath out in an unbroken stream of air; when we think something funny, we let our breath outward, in chunks as it were. Stream of thought and stream of breath follow along the same dynamic line, just as James averred when he wrote that stream of breath accompanies stream of thought. Whether a matter of thought or breath, the stream is clearly a dynamic phenomenon. A qualitative dynamics infuses both inspiration and expiration, and in further ways readily recognizable well over 100 years ago by Sir Charles Bell. In his third edition of *The Anatomy and Philosophy of Expression*, published in 1844, Sir Charles noted that "the first sound of fear is in drawing, not in expelling the breath; for at that instant to depress or contract the chest would be to relax the muscles of the arms and enfeeble their exertion." To make the point more strongly, Sir Charles asks the reader to imagine two men wrestling in the dark, asking whether "the violence of their efforts" would not be apparent from the sounds they make: "The short exclamation choked in the act of exertion, the feeble and stifled sounds of their breathing, would let us know that they turned, and twisted, and were in mortal strife."[6]

James himself captures some aspects of the dynamic relationship between breathing and thinking more closely in his chapter on "The Consciousness of Self." He writes, for example, that:

> In consenting and negating, and in making a mental effort, the movements [felt, he says, inside his head] seem more complex, and I find them harder to describe. The opening and closing of the glottis plays a great part in these operations, and, less distinctly, the movements of the soft palate, etc., shutting off the posterior nares from the mouth. My glottis is like a sensitive valve, intercepting my breath instantaneously at every mental hesitation or felt aversion to the objects of my thought, and as quickly opening, to let the air pass through my throat and nose, the moment the repugnance is overcome. The feeling of the movement of this air is, in me, one strong ingredient of the feeling of assent. [...] In *effort* of any sort, contractions of the jaw-muscles and of those of respiration are added to those of the brow and glottis, and thus the feeling passes out of the head properly so called.[7]

It is notable, however, that even with these fine-grained descriptions of closely attended and exactingly monitored experiences, James misses something basic, something foundational to animate life, namely, kinesthesia, the phenomenon of movement itself,[8] and with it, the kinesthetically-felt qualitative dynamics of self-movement and the kinetic dynamics perceived in the movement of others. That he misses kinesthesia is all the more perplexing in that he implicitly grasps and describes dynamics in what he says and writes of breathing: "breath outward, between the glottis and the nostrils" is an experience of outward *flow*; "[t]he feeling of the movement of this air" is a feeling of *flow*. In short, breath flows and flows with a certain qualitative kinetic or kinesthetically-felt dynamic.

We might note that the oversight might well stem in part at least from a concern with "muscle sense," "the leading organ of space-perception" in the popular view of the day that James rebuts.[9] In his lengthy chapter on "The Perception of Space," he discusses "the muscle sense" and does so from a "muscle contraction" perspective, proceeding to involve himself first in the question of "joint-feelings" and then in the question of "feelings of muscular contraction."[10] Though he does not use such examples in his rebuttal, he would surely agree that when we reach for a glass of water, or walk down the street to a meeting we dread, or throw our arms around a friend, we do not experience joint-feelings or feelings of muscular contraction: we feel our movement—of reaching, walking, or embracing; each of these movements creates a distinctive qualitative dynamic as it unfolds, and in which a distinctive breath-flow, in and out, obtains depending on specific circumstances. When we reach for a glass of water, for example, we are commonly affectively inspired to inspire, savoring its refreshment in advance as it were; when we walk down the street to a meeting we dread, our breath is commonly in rhythm with the dread—short on the inspiration and long on the expiration; when we reach out to embrace a friend, our breath is commonly long on inspiration and long on expiration—long on inspiration as we open our arms and reach out to hug, and long on expiration as we rest in the embrace. Reaching, walking, and embracing are in other words kinetic themes tied intimately with the flow of our breath in and out. They are kinetic themes with multiple possible variations according to the people with whom, or the objects with which we are interacting, and according to the particular situation in which we find ourselves.

In sum, our breath involuntarily modulates its rhythm—its timing, its amplitude, its intensity—to our movement. *Its* movement is an integral part of *our* movement. Indeed, breath is not simply involuntarily coordinated with speaking, for example, with all those fine-grained articulatory gestures by which words flow out of our mouths, it is coordinated with all our bodily movement, and from the beginning of our lives. From this real-time, real-life vantage point, James's insight into the intimate dynamic relationship between thinking and breathing is an insight worthy of wide-ranging recognition, beginning with recognition of the fact that the respiratory center is sequestered in the lowly brain stem—the respiratory center encompasses the medial and lateral reticular fields overlying the rostral half to two-thirds of the inferior olivary nuclei—but is nonetheless powerfully interconnected with all that lies above and below, from frontal lobes to toes. This fact is not likely to surprise Jamesian readers. Hewing to James's extensive research and writings and his seemingly unquenchable inquisitive spirit, readers are led to release themselves from reductionist credos, unleash themselves from ideological tetherings, and investigate on their own. They are furthermore open to the conclusions reached by James after careful weighing of all available evidence, conclusions such as the following appearing at the end of his 68-page chapter on "The Functions of the Brain:" (1) "[A]ll the [nervous] centres, in all animals, whilst they are in one aspect mechanisms, probably are, or at least once were, organs of consciousness in another [aspect], although the consciousness is doubtless much more developed in the hemispheres than it is anywhere else";[11] (2) "All nervous centres have then in the first instance one essential function, that of 'intelligent' action. They feel, prefer one thing to another, and have 'ends.'"[12] As noted at the beginning of this essay, with respect to this latter conclusion especially, well-known physiological psychologist Roger Sperry stressed many times over that perception is not a "mere passage of sensory patterns into passive brain protoplasm"[13] but is "basically an implicit preparation to respond."[14] Indeed, *responsivity* is "a fundamental and almost universal characteristic" of life, of living organisms.[15] Sperry specified this fundamental character of life in finer terms when he underscored the fact that the brain is an organ of and for movement and that the function of consciousness is "*coordinated movement*."[16] James's conclusion that the "one essential function" of "nervous centres" in the brain is "intelligent action" avers just that.

Complementarities and coordination dynamics

James was a dualist, but a dualist unwedded to polarities: "clay and flame"—brain and mind—rather than constituting a dichotomy are inextricably linked. As the quote from James in the opening epigraph affirms, brain and mind "hang indubitably together and determine each other's being." Precisely because they do, they constitute *complementarities* rather than polarities. Moreover the linkage that James points out between thinking and breathing attests to the reality of *coordination dynamics* at all levels of animate life, from the neurological to the behavioral.[17] Studies in coordination dynamics in fact consistently pinpoint and formally articulate neurological and behavioral complementarities, complementarities such as stability~instability,[18] convergence~divergence, central~peripheral, discrete~continuous, and so on. In so doing and in the broadest of terms, they detail the complementary nature of formal elements both to work together and to carry on singly. In the words of J. A. Scott Kelso and David A. Engstrøm, investigations into and studies of coordination dynamics uncover not either-or categories but *complementary pairs* (Kelso and Engstrøm 2006). Notably enough, in their discussion of phase transitions in self-organizing systems and of a metastable dynamics, Kelso and Engstrøm pointedly reference James's comparison of consciousness to the complementary "perchings" and "flights" of a bird.[19]

J. A. Scott Kelso, founder and for 20 years Director of the Center for Complex Systems and Brain Sciences, originated the study of coordination dynamics in conjunction with his investigations and research into brain and behavior.

As with James's writings, Kelso's work too is of pivotal moment in this age of neuroscientific reductionism. Though Kelso and Engstrøm do not mention inspiration and expiration in their two exemplary dictionaries on complementary pairs—the one dictionary organized according to "fields of endeavor," the other specifying prototypical complementary pairings—inspiration and expiration are prime examples of our "complementary nature." The functional division of our autonomic nervous system into sympathetic and parasympathetic nervous systems is a further prime example. These complementarities resonate unequivocally in the course of our everyday lives. The sympathetic system opens respiratory passages; the parasympathetic system constricts them. James's description of his glottis as a "sensitive valve" that modulates his breath in relation to "attending, assenting, negating, making an effort"[20] testifies both to the complementarities inherent in breathing itself, i.e. to inspiration and expiration, and to the complementary pairing of breathing and thinking. His description implicitly testifies as well to the coordination dynamics underlying those complementary pairings, i.e. the coordination dynamics of breathing itself and the relationship of those dynamics to the coordination dynamics of the autonomic nervous system. Moreover, it implicitly testifies to the coordination dynamics of thinking as well as breathing.

Complementarities and the coordination dynamics on which they rest are furthermore apparent affectively. A psychology textbook points out, for example, that "intense pain has profound effects upon the autonomic nervous system, especially its sympathetic branch—heartbeat and respiration are accelerated … Since this pattern of sympathetic arousal usually occurs in situations in which humans experience the emotion fear, it is often called the fear response, whether it occurs in animals or in man."[21] In short, the complementary nature and coordination dynamics of the neurological and the experiential are as livingly present in first-person experiences of emotion as they are in first-person experiences of thinking. James leaves no doubt of this fact in his detailed inquiry into the nature of emotions and his descriptions of feelings of fear, rage, grief, and so on. Of rage, for instance, he writes, "Can one fancy the state of rage and picture no ebullition in the chest, no flushing of the face, no dilatation of the nostrils, no clenching of the teeth, no impulse to vigorous action, but in their stead limp muscles, calm breathing, and a placid face?"[22] Further still, in his chapter on "The Production of Movement," James states, "Using sweeping terms and ignoring exceptions, *we might say that every possible feeling produces a movement, and that the movement is a movement of the entire organism, and of each and all its parts*."[23] This *dynamic congruency*[24] of emotions and movement is of signal evolutionary as well as existential significance: emotions move through the bodies of animate beings and move those bodies to move. What is of further interest in this context is James's alignment of breathing and emotions:

> *The effects upon respiration* of sudden stimuli are also too well known to need elaborate comment. We "catch our breath" at every sudden sound. We "hold our breath" whenever our attention and expectation are strongly engaged, and we sigh when the tension of the situation is relieved. When a fearful object is before us we pant and cannot deeply inspire; when the object makes us angry it is, on the contrary, the act of expiration which is hard.'[25]

James's extensive investigations into what might be termed the "natural science psychology of human being"[26] testify furthermore to the complex nature of complementarities and the coordination dynamics on which they are based, particularly with respect to what in present-day neuroscience is referred to in the one-size-fits-all term: *the brain*. When James writes, "*The 'entire brain process' is not a physical fact at all.* It is the appearance to an onlooking mind of a multitude of physical facts. 'Entire brain' is nothing but our name for the way in which a million of molecules arranged in certain positions may affect our sense."[27] James's empirically-grounded observation presciently echoes that of physicist Niels Bohr with respect to the uncertainty principle. Bohr remarked upon "*the impossibility of any sharp separation between the behavior of atomic objects and the interaction with the measuring instruments which serve to define the conditions under which the phenomena appear*."[28] Ignoring such empirically certain observations, many "onlooking minds" in present-day neuroscience unashamedly declare "facts" about "the entire brain." Consider, for example, the following advertisement by The Teaching Company[29] that specifies a course taught by a neuroscience professor at a well-known and reputable university; the course is titled "How Your Brain Works." It is described as follows: "Everything you hear, feel, see, and think is controlled by your brain. It allows you to cope masterfully with your everyday environment and is capable of producing breath-taking athletic feats, sublime works of art, and profound scientific insights. But its most amazing achievement may be that it can understand itself."[30] In such "onlooking mind" courses and in such "onlooking mind" articles and texts, *the brain* is clearly the oracle at Delphi, the Mecca, so to speak, to which all questions about ourselves and our relationship to our surrounding world are addressed and from which all questions will in time be duly answered. In effect, the *fons et origo* of our effective relationships with other humans and with the array of objects we experience in the world about us, the fount of our knowledge of the world and of other people is neurological and lies in our own heads in the form of "internal non-linguistic 'representations' of the body-states associated with actions, emotions, and sensations." In the summary announcement of his guest lecture in conjunction with his Arnold Pfeffer 2009 prize award, Vittorio Gallese states just that.[31]

Clearly, in following along reductionist lines, we are a long way from inspiration and expiration, the autonomic nervous system, coordination dynamics, and complementarities. Moreover disregarding James's admonishment that "if a brain *could* be self-conscious, it would be conscious of itself *as* a brain and not as something of an altogether different kind," many "onlooking minds" in present-day neuroscience, as noted earlier, think nothing of making experiential ascriptions to the brain, precisely as in the above citation that not only has the brain producing "breathtaking athletic feats, sublime works of art, and profound scientific insights," and also unequivocally state that the brain's "most amazing achievement may be that it can understand itself." At the very beginning of a chapter titled "What Is It Like To Be a Brain,"[32] I did not cite James's insightful observation, but I did point out that "to be a brain is to be unilluminated"; in other words, brains are *inside* and thus in the dark, and that, moreover, brains "never show themselves directly except by way of surgery or accidents, or by way of dissections; they show themselves indirectly only through experimentation." As the chapter goes on to point out, however, "brains are considered to be *normally active*, and to be so continuously until the moment the organism in which the brain inheres, dies."[33]

Research findings in coordination dynamics attest to that "considered" fact: "behaviors such as perceiving, intending, acting, learning, and remembering arise as metastable spatio-temporal patterns of brain activity that are themselves produced by cooperative interactions among neural clusters. Self-organization is the key principle."[34] A key research principle affirms that key dynamic principle of self-organization: one "look[s] beyond the single cell level, to try to find relevant variables that characterize the large-scale cooperative activity of the human brain."[35] The notion of "cooperative activity" echoes James's conclusion that the "one essential function" of "nervous centres" in the brain is "intelligent action." "Nervous centres" clearly work together toward a common end. Moreover, the notion echoes James's later "confession" regarding the integrity of the brain and the relationship between thoughts or consciousness and the brain. James states first that "the whole brain must act together if certain thoughts are to occur," and then goes on to state:

> The consciousness, *which is itself an integral thing not made of parts*, "corresponds" to the entire activity of the brain, whatever that may be, at the moment. This is a way of expressing the relation of mind and brain from which I shall not depart during the remainder of the book, because it expresses the bare phenomenal fact with no hypothesis, and is exposed to no such logical objections as we have found to cling to the theory of ideas in combination.[36]

Polarities and reductionism

The propensity to atomize goes back a long, long time as G. E. R. Lloyd points out in his book *Polarity and Analogy: Two Types of Argumentation in Early Greek Thought*.[37] After noting that "Many factors appear to contribute to the remarkable prevalence of theories based on opposites in so many societies at different stages of technological development," he notes that:

> First there is the fact that many prominent phenomena in nature exhibit a certain duality: day alternates with night; the sun rises in one quarter of the sky and sets in the opposite quarter; in most climates the contrast between seasons (summer and winter, or dry season and rainy season) is marked; in the larger animals male and female are distinct, and the bilateral symmetry of their bodies is obvious.[38]

He goes on to cite two further factors: (1) "the duality of nature often acquires an added significance as the symbolic manifestation of fundamental religious or spiritual categories: the classification of phenomena into opposite groups may reflect, and itself form an important part of, a system of religious beliefs which expresses the ideals of the society, and by which the whole life of the society is regulated;" and (2) "whether or not the terms are divided into a 'positive' and a 'negative' pole, opposites provide a simple framework of reference by means of which complex phenomena of all sorts may be described or classified." His conclusion: "Antithesis is an element in any classification, and the primary form of antithesis, one may say, is division into *two* groups—so that the *simplest* form of classification, by the same token, is a dualist one."[39]

However thoughtfully rigorous and finely illuminating Lloyd's analysis of theories based on opposites, as James's detailed inquiry into mind and brain shows, dualism need not necessarily result in polarities; that is, a dualistic relationship does not automatically require a "theory of opposites," though it might be noted that a theory of opposites might for some individuals reside in a conceptual disposition toward binary oppositions.[40] Neither need a reductionist stance be adopted in order to dissolve a dualism, as James's reflections affirm. Finally too, neither need solutions to the "mystery" of dualism be invoked to soften the gap or impasse. While solutions to the "mystery" of mind might be offered—for example, the solution of a "natural teleology"[41]—it is notable that James was content to live with "mystery." As the opening epigraph attests, "the two things [mind and brain] hang indubitably together and determine each other's being, but how or why, no mortal may ever know."

Yet it is also possible to acquaint oneself more closely with life in relation to the mind-brain dualism and in particular *with the evolution of animate forms of life*, investigating and inquiring precisely in the manner of William James, whose chapter on "The Mind-Stuff Theory" ends with the epigraph that introduced this essay. James in fact begins that chapter with remarks about evolution. He points out first that:

> As long as we keep to the consideration of outward facts, even the most complicated facts of biology, our task as evolutionists is comparatively easy. We are dealing all the time with matter and its aggregations and separations; and although our treatment must perforce be hypothetical, this does not prevent it from being *continuous*. The point which as evolutionists we are bound to hold fast to is that all the new forms of being that make their appearance are really nothing more than results of the redistribution of the original and unchanging materials. … In this story no new *natures*, no factors not present at the beginning, are introduced at any later stage.

As James immediately notes, however, "But with the dawn of consciousness an entirely new nature seems to slip in, something whereof the potency was *not* given in the mere outward atoms of the original chaos." He then observes that "The enemies of evolution have been quick to pounce upon this undeniable discontinuity in the data of the world, and many of them, from the failure of evolutionary explanations at this point, have inferred their general incapacity all along the line."[42] Though he could certainly not be described as an enemy of evolution, philosopher Thomas Nagel's negative view of materialist neo-Darwinians' explanations of mind or consciousness[43] is a recent instance of this "pouncing," a definitively thoughtful and explorative kind of pouncing rather than "quick," but one no less concerned with "an entirely new nature slip[ping] in."[44]

In contrast to a recognition and investigation along evolutionary lines and to solutions offered by "enemies of evolution," many if not most present-day neuroscientists, cognitive scientists, and cognitive philosophers follow a singular and unwavering reductive path, for such a path presents itself to them as a clear and clean way out. In fact, to judge from the Teaching Company advertisement, reductionism enters effortlessly to solve the impasse. At the same time that it is introduced, however, it commonly blocks further possible investigations and inquiries, swiftly focusing on *the brain*, the incomparable human one, and a thoroughly generic incomparable human one at that, as indicated in the quotation from the Teaching Company. An accommodating answer to the extremism that basically characterizes reductionist thinking and programs is readily available. In the abstract of their article "Outline of a General Theory of Behavior and Brain Coordination," Kelso and his co-authors Guillaume Dumas and Emmanuelle Tognoli state:

> We report a program of research that uses the theoretical concepts of coordination dynamics and quantitative measurements of simple, well-defined experimental model systems to explicitly relate neural and behavioral levels of description in human beings. Our approach is both top-down and bottom-up and aims at ending up in the same place: top-down to derive behavioral patterns from neural fields, and bottom-up to generate neural field patterns from bidirectional coupling between astrocytes and neurons.

They then note, "Much progress can be made by recognizing that the two approaches—reductionism and emergentism—are complementary. A key to understanding is to couch the coordination of very different things—from molecules to thoughts—in the common language of coordination dynamics."[45]

Just such a research program brings us back again to James's holistic view of mind and matter. In fact, in *The Complementary Nature*, Kelso and Engstrøm cite James's memorable description of consciousness as comparable to the perchings and flights of birds. They point out that "James's beautiful metaphor of perchings and flights for the stream of consciousness is realized physically by metastable coordination dynamics. His perchings correspond to integrative, phase-gathering tendencies … and his flights, the drift between 'perching' plateaus … to the segregative, phase-scattering escapes that explore the space of possibilities." Moreover Kelso and Engstrøm go on to state:

> *Thinking* … is a transient nonstationary dynamic process. It corresponds to a flow of converging "perchings" (integrative phase-locking tendencies between brain areas) and diverging "flights" (segregative decoupling tendencies and individuation of brain areas). Both tendencies are crucial: the former to create thoughts, feelings … the latter to release individual brain areas to participate in other acts of cognition and emotion.[46]

It is of interest to note further that their final sentence in the paragraph highlights the hazards of tethering oneself to a choice of either reductionism or dualism—or, we might add, a solution to "the mystery" such as "natural teleology." Kelso and Engstrøm write, "To be stuck in a phase-locked state is to be temporarily 'trapped in thought', to be depressed in affect, in one stationary state or another, the limited repertoire of the either/or."[47]

Dynamics, evolution, whole-body organisms, and experience: Concluding reflections

The "unstable equilibrium" of instinct of which James writes is a prescient validation of what Kelso and Engstrøm describe as the process of thinking, namely, that thinking, with its "perchings" and "flights," is, in Kelso and Engstrøm's terms, a "metastable phenomenon," in James's terms, a "nonstationary dynamic process." In particular, James observes that "greediness and suspicion, curiosity and timidity, coyness and desire, bashfulness and vanity, sociability and pugnacity, seem to shoot over into each other as quickly, and in an unstable equilibrium, in the higher birds and mammals as in man."[48] The contraries that James describes as "an unstable equilibrium" might be understood as polarities or as complementary pairs that undergird the intrinsic dynamics of animate life, those intrinsic dynamics being the cornerstone of coordinated movement across the spectrum of animate life. As Kelso and Engstrøm remark, for example, "a fetus moves spontaneously within the womb, ... discover[ing] arms and legs that flex~extend, a mouth that opens~closes, a body that bends~twists."[49] In like fashion, James remarks that:

> Since what seems to be the same object may be now a genuine food and now a bait; since in gregarious species each individual may prove to be either the friend or the rival, according to the circumstances, of another; since any entirely unknown object may be fraught with weal or woe, *Nature implants contrary impulses to act on many classes of things*, and leaves it to slight alterations in the conditions of the individual case to decide which impulse shall carry the day.[50]

What is furthermore of interest in the way of coincidence is the way in which James implicitly harmonizes contraries into complementarities when he states that impulses are superseded by choice on the basis of experience: "*The animal that exhibits them loses the 'instinctive' demeanor* and appears to lead a life of hesitation and choice, an intellectual life."[51] Moreover he goes on to speak of impulses in relation to reason, observing that "*man possesses all the impulses that they* ['lower creatures'] *have, and a great many more besides*" and that "[i]n other words, there is no material antagonism between instinct and reason."[52]

What is of moment too in this discussion of the "unstable" or "metastable" dynamics of animate life is precisely an evolutionary perspective: from whatever academic perspective we might view them, James's principles of psychology pertain not to humans alone but derive from the evolution of animate life. When we in fact look across the spectrum of animate life, we perforce are witness to, and experience first-hand, a diversity of ways of making a living, ways that are tied to particular morphologies. Indeed, we witness *morphologies-in-motion*.[53] This reality accords with James's observation that impulses generated and permeated by instincts propel animals, including humans, toward certain kinds of movement in their everyday lives, movement that bespeaks curiosity or timidity, sociability or pugnacity, and so on. It is thus hardly surprising that when we look across the spectrum of animate life, we see synergies of meaningful movement:[54] *synergies of meaningful movement, intelligent action, coordination dynamics, coordinated movement, responsivity*—all attest to the same bio-existential fact of life, namely, a kinetic dynamics itself founded on the build-up of tactile-kinesthetically engendered[55] kinetic/affective habits undergirding the basic animate phenomenon of responsivity[56] and the familiarity of the everyday world.[57] In this context, it is sobering to recall Darwin's observation: "Experience shows the problem of the mind cannot be solved by attacking the citadel itself—the mind is function of body—we must bring some *stable* foundation to argue from."[58] Though not in answer to Darwin's insight and urging, present-day tomes such as *How the Body Shapes the Mind*,[59] *The Phenomenological Mind*[60] and *The Body in the Mind*[61] are testimonials to the effort to heed the insight and answer to the urging. Yet however bodily-tethered such efforts, the *stable* foundation that Darwin urges eludes them. That stable, *foundational* level of bodily life in fact remains unreached in much if not most of present-day literature on mind and body. Perhaps that is no wonder since the *stable* foundation is, ironically, the ineluctable modality of movement that pervades animate life. James's short but singularly insightful chapter on "The Production of Movement" affirms just that in its very beginning observation: "Using sweeping terms and ignoring exceptions, *we might say that every possible feeling produces a movement, and that the movement is a movement of the entire organism, and of each and all of its parts*."[62] Throughout this brief ten-page chapter, James in fact emphasizes over and over the kinetic involvement of the whole body. In summing up in retrospect what he has written, he states, "Looking back over all these facts, it is hard to doubt the truth of the law of diffusion, even where verification is beyond reach. *A process set up anywhere in the centres reverberates everywhere, and in some way or other affects the organism throughout, making its activities either greater or less*."[63]

Given these several vantage points—of dynamics, evolution, and whole-body organisms—it is difficult to maintain a reductionist brain-centered perspective as in much of present-day neuroscience, cognitive science, and cognitive philosophy. Indeed, James's retrospective summation on what we might call the whole animate body readily recalls his descriptions of the foundational reality of breathing. Due and rigorous attention to that foundational reality might readily induce reductionists to turn their attention downward, to the body as a whole, starting perhaps even with an attention to the lowest major portion of "the brain," the brain stem, which regulates essential physiological processes, and, we might note, not only respiration, but wake-sleep cycles, heart rate, and blood pressure. They might thus awaken to the truth that James points out, namely, that "the whole brain must act together if certain thoughts are to occur."[64] James in fact goes on to affirm that "The consciousness, which is itself an integral thing not made of parts, 'corresponds' to the entire activity of the brain, whatever that may be, at the moment" and to warn readers in advance that "This is a way of expressing the relation of mind and brain from which I shall not depart during the remainder of the book, because it expresses the bare phenomenal fact with no hypothesis, and is exposed to no such logical objections as we have found to cling to the theory of ideas in combination."[65]

An awakened attention to breathing might furthermore awaken the realization that such attention requires listening to the body. A fitting acknowledgment of the legacy of William James would be to listen to the body as James does virtually throughout his principles of psychology. Though the specter of introspection might be raised, it is equally fitting that it be lowered in recognition of the fact that *experience*, including experience of the pan-animate phenomenon of breathing, can hardly be discounted as irrelevant to the realities of mind and body and to our knowledge of those realities. We need only recall that neurons develop and parcel out in the course of infancy. In their recent second edition of *Cognition, Brain, and Consciousness: Introduction to Cognitive Neuroscience*, in which they cite Nobel prize-winning neuroscientist Gerald Edelman's research on neuronal development, cognitive neuroscientists Bernard J. Baars and Nicole M. Gage state, "[T]ens of billions of neurons sprout as the fetus grows, each new cell following its own pathway to make up the brain, spinal cord, and peripheral neurons," "peripheral neurons" undoubtedly referring to all neurons throughout the living body, some of which connect to and from neurons in the brain and spinal cord, neurons such as those found in "human biological motion," i.e. neurons in muscles, tendons, and joints.[66] They point out that "when a baby is born, it will have more nerve cells than it will ever again have in its entire lifetime" and this because "neurons are constantly being *pruned*, if they make unstable or unproductive connections."[67] Moreover, Baars and Gage point out that "Once the fetus begins to use its senses and learn, around the beginning of the third trimester of pregnancy, the production of new neurons slows down, but "new *synaptic connections* keep sprouting in the trillions."[68] It warrants emphasis that Baars and Gage speak of a fetus beginning to *sense* and *learn*.[69] Baars and Gage conclude by stating that "The connectivity pattern of the brain is the key to its working" and that "synaptic growth and selective pruning continues throughout life."[70] As noted elsewhere,[71] their later observation that "Brain development begins with zero synapses and ends with a thousand trillion connections"[72] brings strikingly to the fore the fundamental import of connectivity and pruning, and with it, not just the unmistakable import of interneuronal dynamic patternings in *the brain*, but *the unmistakable import of animate life itself in the form of experience* that shapes connectivity and pruning, experience in which movement—animation—plays a quintessentially central role.[73]

The research conclusions of neuroscientist Gerald Edelman are equally striking. In the context of recognizing the fact that "[t]he network of the brain is created by cellular movement during development and by the extension and connection of increasing numbers of neurons,"[74] Edelman points out with respect to pruning that "in some regions of the developing nervous system up to 70 percent of the neurons die before the structure of that region is completed!" and that "[i]n general, … uniquely specified connections cannot exist."[75] After pointing out that the cerebral cortex houses ten billion neurons—not counting glia cells—and that "[e]ach nerve cell receives connections from other nerve cells at sites called synapses," he comments, "But here is an astonishing fact—there are about one million billion connections in the cortical sheet. If you were to count them, one connection (or synapse) per second, you would finish counting some thirty-two million years after you began."[76] He goes on to point out that "[i]f we consider how connections might be variously combined, the number would be hyperastronomical—on the order of ten followed by millions of zeros."[77] Moreover, a further observation of his concerns precisely the absence of genetic determinism: synaptic connections among neurons are not prespecified in a definitive way by our genes.[78] The brain is indeed a *dynamic self-organizing system*, as Kelso and his colleagues have shown.[79] In short, the enormous sprouting of neuronal cells and their pruning in the course of fetal development, their synaptic growth and pruning in the course of life, and the sheer number of possible synaptic connections that might be made attest without question to the brain's foundational internal dynamics. Thus, however spatially fixed and permanent the anatomical parts of the brain, its neurology in a living sense in a living being is on the move, unbound to spatially fixed and permanent pathways.

Given the above neurological facts from research findings, we might dolefully exclaim, "Ah! That present-day neuroscientists, cognitive scientists and philosophers would listen to their bodies as James did!" Rather than unabashedly or unthinkingly purloining real-time, real-life experience by spurious experiential ascriptions to *the brain*, such listening might attune researchers to the myriad ways in which experience and the dynamics of animate life itself shape "the brain." At the very least, such listening might prompt reductionists to acknowledge that they are indeed "onlookers," that the brain at which they are looking is "the appearance to an onlooking mind of a multitude of physical facts" and that even the "'[e]ntire brain' is nothing but our name for the way in which a million of molecules arranged in certain positions may affect our sense"—and, we might add, our feelings.

Notes

1 James 1950 [1890], vol. 1, p. 182.
2 *Ibid.*, p. 343.
3 *Ibid.*, p. 79.
4 James 1987 [1904], pp. 1157–8.
5 Jung 1968 [2nd edn], pp. 8–9; see also Sheets-Johnstone 2008, pp. 357–8 for implications with respect to movement in relation to emotions.
6 Bell 1844, pp. 190–1.
7 James 1950 [1890], vol. 1, p. 301.
8 E.g. James 1950 [1890], vol. 2, p. 190, in his chapter on "The Perception of Space."
9 *Ibid.*, pp. 197–202, see especially p. 202.
10 *Ibid.*, pp. 134–282.
11 James 1950 [1890], vol. 1, p. 78.
12 *Ibid.*, p. 79.
13 Sperry 1952, p. 302.
14 *Ibid.*
15 Curtis 1975, p. 28.

16 Sperry 1939, p. 295; italics added.
17 Kelso 1995; Kelso 2009.
18 Kelso and Engstrøm use a tilde or "squiggle mark" to identify complementary pairs. As they explain and go on to demonstrate, "We use the tilde ~ not to concatenate words or as an iconic bridge between polarized aspects, but to signify that we are discussing complementary pairs. Equally if not more important, the tilde symbolizes the *dynamic nature* of complementary pairs" (Kelso and Engstrøm 2006, p. 7).
19 *Ibid.*, pp. 175–6.
20 James 1950 [1890], vol. 1, pp. 300–1.
21 Gleitman 1981, p. 143.
22 James 1950 [1890], vol. 2, p. 452.
23 James 1950 [1890], vol. 1, p. 372.
24 Sheets-Johnstone 1999a.
25 James 1950 [1890], vol. 2, p. 376.
26 James 1950 [1890], vol. 1, p. v.
27 *Ibid.*, vol. 1, p. 178.
28 Bohr 1949, p. 210: N. Bohr (1949) "Discussion with Einstein on Epistemological Problems in Atomic Physics," in P. Schlipp, ed. *Albert Einstein: Philosopher-Scientist*, La Salle: Open Court.
29 *Science News*, vol. 175, no. 13, p. 3.
30 *Ibid.*
31 Galleses 2010, "From Mirror Neurons to Embodied Simulation." Available at: http://npsafoundation.org/PfefferCenter/Lectures/page15/opage15.html
32 Sheets-Johnstone 1999b/2011.
33 *Ibid.*, pp. 453–4.
34 Kelso 1995, p. 257.
35 *Ibid.*, pp. 257–8.
36 James 1950 [1890], vol. 1, p. 177; italics added.
37 Lloyd 1966.
38 *Ibid.*, p. 80.
39 *Ibid.*
40 Sheets-Johnstone 2000.
41 Nagel 2012.
42 James 1950 [1890], vol. 1, p. 146.
43 Nagel 2012.
44 For a refutation of James's claim of an "undeniable discontinuity in the data of the world" (and of Nagel's claim of same), see Sheets-Johnstone 1999a/expanded 2nd edn, 2011, Chapter II, Part I, "Consciousness: A Natural History." The chapter was originally published in the *Journal of Consciousness Studies* 5: 3 (1998), pp. 260–94.
45 Kelso et al. 2013, p. 120.

46 Kelso and Engstrøm 2006, p. 176.
47 *Ibid.*
48 James 1950 [1890], vol. 2, p. 392.
49 Kelso and Engstrøm 2006, pp. 105–6.
50 James 1950 [1890], vol. 2, p. 392; italics in original.
51 *Ibid.*, pp. 392–3.
52 *Ibid.*, p. 393; italics in original.
53 Sheets-Johnstone 2010a, 2010b, 2011a.
54 Sheets-Johnstone 2011b.
55 Or tactile-proprioceptively engendered: see Sheets-Johnstone 1999b/2011.
56 Curtis 1975.
57 *Ibid.*
58 Darwin 1987 [1836–44], Notebook N, p. 564, #5.
59 Gallagher 2005.
60 Gallagher and Zahavi 2012.
61 Johnson 1987.
62 James 1950 [1890], vol. 2, p. 372.
63 *Ibid.*, p. 381.
64 James 1950 [1890], vol. 1, p. 177.
65 *Ibid.*
66 Baars and Gage 2010, pp. 516, 171, respectively.
67 *Ibid.*, p. 516.
68 *Ibid.*; italics in original.
69 See also Furuhjelm et al. 1976, on the earlier movement capabilities of fetuses, movement capabilities such as kicking, sucking, waving, and turning around; Windle 1971, on the prenatal development of kinesthesia, in particular, on neural receptors in muscles that provide a sense of movement and position; and Piontelli's comments (1987, pp. 455, 456) on "the richness and the complexity of movements one could observe [in the fetus] since the early stages" and on "the freedom of movement each foetus could enjoy in the amniotic fluid."
70 *Ibid.*
71 Sheets-Johnstone 2012.
72 Baars and Gage 2010, p. 217.
73 Obviously, fetal sensing and learning center quintessentially on movement. But obviously too, that sensing and learning continue when a fetus is actually born: *it comes into the world moving*. It is not only not stillborn, but continues to move and, in the process, learns on the basis of inchoate reaching movements to reach effectively and to grasp whatever object is attracting it, for example. It learns to turn over in its crib; it learns to crawl and to walk. In the most basic sense, it learns to move effectively and efficiently in its surrounding world. Moreover, as Baars and Gage point out, synaptic growth and selective pruning is a life-long process, a

process that, as concerns movement, depends precisely upon the degree to which movement informs one's individual life and the degree to which it is curtailed, that is, the degree to which one suffers "[t]he heart-ache and the thousand natural shocks that flesh is heir to" (Shakespeare, *Hamlet*, Act III, Scene 1).

74 Edelman 1992, p. 25.

75 *Ibid.*, p. 17.

76 *Ibid.*

77 *Ibid.*

78 *Ibid.*, p. 23.

79 E.g. Kelso 1995; Kelso 2009; Kelso and Engstrøm 2006; Jantzen et al. 2008; Kelso et al. 2013.

References

Baars, B. J., and Gage, N. M. (2010), *Cognition, Brain, and Consciousness: Introduction to Cognitive Neuroscience*, 2nd edn. Amsterdam: Elsevier.

Bell, Sir Charles (1844), *The Anatomy and Philosophy of Expression.* London: John Murray.

Control of Breath Website: www.physioweb.org/respiration/control_breath.html

Curtis, Helena (1975), *Biology*, 2nd edn. New York: Worth Publishers.

Darwin, Charles (1987), *Charles Darwin Notebooks, 1836–1844*, Paul H. Barrett, Peter J. Gautrey, Sandra Herbert, David Kohn, Sydney Smith (eds). Ithaca, NY: Cornell University Press.

Edelman, G. (1992), *Bright Air, Brilliant Fire: On the Matter of the Mind.* New York: Basic Books.

Furuhjelm, M., Ingelman-Sundbert, A., and Wirsén, C. (1977), *A Child Is Born*, rev. edn. New York: Delacourte Press.

Gallagher, Shaun (2005), *How the Body Shapes the Mind.* Oxford: Clarendon Press/ Oxford University Press.

Gallagher, Shaun and Zahavi, Dan (2012), *The Phenomenological Mind*, 2nd edn. London and New York: Routledge.

Gleitman, Henry (1981), *Psychology.* New York: W. W. Norton & Company.

James, William (1950 [1890]), *The Principles of Psychology*, vols 1 and 2. New York: Dover Publications.

—(1987 [1904]), "Does 'Consciousness' Exist?" in *William James: Writings 1902–1910.*" New York: Library of America.

Jantzen, Kelly J., Steinberg, Fred L., and Kelso, J. A. Scott (2008), "Coordination Dynamics of Large-scale Neural Circuitry Underlying Rhythmic Sensorimotor Behavior," *Journal of Cognitive Neuroscience* 21(12): 2420–33.

Johnson, Mark (1987), *The Body in the Mind.* Chicago, IL: University of Chicago Press.

Jung, Carl G. (1968), *Aion: Researches into the Phenomenology of the Self*, trans. R. F. C. Hull. Bollingen Series XX. Princeton, NJ: Princeton University Press.

Kelso, J. A. Scott (1995), *Dynamic Patterns: The Self-Organization of Brain and Behavior*. Cambridge, MA: MIT Press/Bradford Books.

—(2009), "Coordination Dynamics," in *Encyclopedia of Complexity and Systems Sciences*, ed. R. A. Meyers. Berlin: Springer, pp. 1537–64.

Kelso, J. A. Scott and David A. Engstrøm (2006), *The Complementary Nature*. Cambridge, MA: MIT Press/Bradford Books.

Kelso, J. A. Scott, Dumas, Guillaume, and Tognoli, Emmanuelle (2013), "Outline of a General Theory of Behavior and Brain Coordination," *Neural Networks* 37: 120–31.

Lloyd, G. E. R (1966), *Polarity and Analogy: Two Types of Argumentation in Early Greek Thought*. Indianapolis, IN and Cambridge, MA: Hackett Publishing.

Nagel, Thomas (2012), *Mind and Cosmos: Why the Materialist Neo-Darwinian Conception of Nature Is Almost Certainly False*. Oxford: Oxford University Press.

Piontelli, A. (1987), "Infant Observation From Before Birth," *The International Journal of Psychoanalysis* 68: 453–63.

Sheets-Johnstone, Maxine (1999a), "Emotions and Movement: A beginning empirical-phenomenological analysis of their relationship," *Journal of Consciousness Studies* 6(11–12): 259–77.

—(1999b/expanded 2nd edn 2011), *The Primacy of Movement*. Amsterdam and Philadelphia, PA: John Benjamins.

—(2000), "Binary Opposition as an Ordering Principle in (Male?) Human Thought," in *Feminist Phenomenology*, Linda Fisher and Lester Embree (eds). Dordrecht: Kluwer Academic Publishers, pp. 173–94.

—(2008), *The Roots of Morality*. College Park, PA: Pennsylvania State University Press.

—(2010a), "Why Is Movement Therapeutic?" *American Journal of Dance Therapy*, 32(1): 2–15.

—(2010b), "Kinesthetic Experience: Understanding Movement Inside and Out," *Body, Movement and Dance in Psychotherapy*, 5(2): 111–27.

—(2011a), "The Imaginative consciousness of Movement: Linear Quality, Kinaesthesia, Language and Life," in *Redrawing Anthropology: Materials, Movements, Lines*, ed. Tim Ingold. Surrey: Ashgate.

—(2011b), "Embodied Minds or Mindful Bodies? A Question of Fundamental, Inherently Inter-related Aspects of Animation," *Subjectivity* 4(4): 451–66.

—(2012), "Movement and Mirror Neurons: A challenging and choice conversation," *Phenomenology and the Cognitive Sciences* 11(3): 385–401.

Sperry, Roger (1939), "Action Current Study in Movement Coordination," *Journal of General Psychology* 20: 295–313.

—(1952), "Neurology and the Mind/Brain Problem," *American Scientist* 40: 291–312.

Windle, W. F. (1971), *Physiology of the Fetus*. Springfield, MA: Charles C. Thomas.

4

Dewey, Naturalism, and Neuroaesthetics

Russell Pryba

Introduction

In an article in *Science* in 2001, Semir Zeki, one of the founders of neuroaesthetics, expressed his vision for the newly emerging field:

> The future of what I call neuroaesthetics will, I hope, study the neural basis of artistic creativity and achievement, starting with the elementary perceptual process. I am convinced that there can be no satisfactory theory of aesthetics that is not neurobiologically based. All human activity is ultimately a product of the organization of our brains, and subject to its laws.[1]

While the first part of this passage lays out the areas of inquiry of a new scientific sub-discipline, the latter portion of the quote moves beyond demarcating the limits of neuroaesthetics to making pronouncements about what constitutes an adequate aesthetics. Zeki's view implies that the only adequate theory of aesthetics, in virtue of the fact that all human activity is the product of brain activity, is neuroaesthetics. What it means to claim that artistic production and aesthetic enjoyment are "ultimately a product of the organization of our brains" is however, significantly less clear and more controversial than merely defining the scope of neuroaesthetics itself. While there can be no doubt that the brain underlies the impressive range of human capabilities and behavior, does it follow that the only adequate understanding of the arts is one that reduces the complex cultural matrix of artistic production and consumption to the electrochemical happenings at the neurological level?

The questions posed by Zeki's statement of the adequacy of an aesthetic theory have implications that range beyond the understanding of the complex relationship between science and the arts. For philosophers interested in developing a comprehensive philosophical naturalism, the claims of neuroaesthetics present an important challenge for assessing the sufficiency of various forms of naturalism. Since the arts have traditionally been thought of as the area of humanistic inquiry that most decisively refuse to yield to a comprehensive analysis in terms of scientific understanding alone, the relationship between neuroaesthetics and philosophical aesthetics represents an important local site in the debate about what constitutes an adequate philosophical naturalism.

In this chapter, I shall address the relationship between philosophical and neuroscientific approaches to the study of the arts and aesthetics by utilizing John Dewey's theories of scientific inquiry and aesthetic experience. A Deweyan approach to the question of the connections between neuroaesthetics and traditional aesthetic theorizing is appropriate for two reasons. First, given Dewey's life-long promotion of scientific method, it is consistent with Dewey's overall philosophical project to pursue a deeper scientific understanding of what underlies aesthetic experience. Secondly, Dewey's insistence that aesthetic experience is continuous with the interactions between an organism and its environment that constitutes experience writ large, illustrates how a scientific approach to understanding the arts would contribute to an understanding of human cognition and experience generally. In other words, since Dewey defended a naturalistic theory of aesthetics that recognizes the continuity between the biological and the aesthetic, his theory is particularly well suited to assess the project of neuroaesthetics from a philosophical perspective. What will emerge from this discussion is the understanding that insofar as neuroaesthetics is conceived as a reductive science it will fail to be a replacement for philosophical approaches to aesthetics which recognize the deeply cultural nature of the arts. Neuroaesthetics will be invaluable to the extent that it is in a unique position to investigate the neural correlates of artistic creativity and aesthetic appreciation.

However, reducing the complex cultural matrix of experience in which our engagement with the arts take place, and in which they must finally be understood, to the brain activity that underlies such phenomena is to rob our explanation and understanding of arts of their central function. Aesthetic experience is valuable insofar as it possesses the ability to enrich human culture and experience as it is lived and to bring this experience to a heightened state of awareness and enjoyment. Insofar as understanding the neurological and biological capacities that allow the arts to achieve this happy issue will succeed in making the meaning of aesthetic experience more deeply qualitatively felt, then neuroaesthetics should be seen as a welcome contributor to the investigation and understanding of one of the most essential and meaningful human activities—the making and appreciating of art.

Not only will this chapter offer a pragmatic framework from which to understand the contribution that neuroaesthetics can make to the overall understanding of the aesthetic dimension of human experience, but in the process it shall also provide an explication of the relationship between art and science in Dewey's thinking. A hallmark of John Dewey's aesthetic theory is that art is an immediately enjoyable meaning. Unlike science, which for Dewey is a second-order reflection upon the doings and sufferings of ordinary experience as it is lived, art need not be mediated by reflective thought in order for it to issue in the consummatory experience that is characteristic of Dewey's analysis of the aesthetic. As a consequence, it might be thought that there is a tension between art and science in Dewey's philosophy as a whole. Further, it could be asserted that Dewey, the champion of the scientific method as the preeminent method for transforming nature in the raw into useful instrumentalities, viewed philosophy as more continuous with science than with the arts. To put the claim another way, it could be argued that there is a tension in Dewey's thought between his commitment to naturalism and his insistence that art is experience.[2] The issue is further complicated by the diversity of philosophical positions that have been referred to as naturalism.

Historically, philosophical naturalism has been understood as the conjunction of two claims: (1) the rejection of the supernatural and (2) the acceptance of the scientific method as the preeminent method for coming to know the nature of reality. While the first is a metaphysical claim and the second epistemological or methodological, there is no clear distinction between the methodological commitment to scientific inquiry expressed in (2) above and the ontological implication of such an acceptance. If it is claimed that the scientific method is the only secure method through which we can come to know the contents of reality, then what is really real—all things considered—are the entities that scientific experimentation discloses. In philosophical terms, the triumph of naturalism has coincided with the rise of reductive physicalism, and recently philosophers have turned more and more to science, especially neuroscience, for fresh perspectives on seemingly intractable philosophical problems.

Yet, a closer examination of Dewey's own naturalistic commitments will show that not only is his advocacy of the scientific method consistent with his aesthetic theory but also that his non-reductive naturalism is a helpful framework in which to understand the relationship between art and science more generally. Given the recent emergence of the field of neuroaesthetics, due in part to the refinement of neural imaging technologies, investigating the relationship between scientific approaches to the arts and more traditional forms of aesthetic theorizing is more urgent than at previous moments in the history of aesthetics.

Rather than thinking of naturalism as a thesis about how philosophy can be more continuous with science, this chapter will present a type of naturalism that enables a deeper understanding of art as it is embedded in the daily cultural practices of living that enrich human experience with immediately perceptible meanings. Thus if, as Dewey insists, science *is* an art, then there can be no fundamental tension between art and science. Further, neuroaesthetics should not be understood as being in a position to resolve fundamental questions about the meaning and value of art as it is embodied in living practice, by replacing culturally informed analyses of the arts for an analysis of the neural, chemical, and physical correlates of aesthetic experience.

Since for Dewey art and the aesthetic are irreducibly cultural phenomena, and the biological is completely continuous with the cultural life of the human organism as it is lived, the biological understanding of art and the aesthetic are an essential component of understanding aesthetic experience as a whole. However, when the biological is taken to be foundational of the aesthetic in such a way that cannot account for how aesthetic experience is qualitatively *felt and enjoyed* as an emergent phenomenon, then the individualizing character of aesthetic experience that Dewey took great pains to emphasize is lost. Investigating the extent to which neuroaesthetics is committed to this reductive project, and thus in tension with Dewey's naturalistic account of the aesthetic, is the primary purpose of this chapter.

In the first section of this chapter, I shall provide an explication of John Dewey's pragmatic non-reductive naturalism—the so-called Columbia Naturalism—through an analysis of the writings of Dewey and his followers. The second section shall address the consistency of Dewey's defense of the scientific method as constitutive of pragmatic naturalism and his aesthetic theory in order to illustrate the continuity of science and the arts in Dewey's thought. I shall then examine the field of neuroaesthetics in order to assess in what ways it is consistent or in conflict with Dewey's understanding of the relationship between art and science. Here I shall not be primarily concerned with specific neuroscientific experiments and results that pertain to an understanding of art and the aesthetic. Rather I shall focus more narrowly on the way in which neuroscientific practitioners understand the scope and range of their project. Thus the section on neuroaesthetics might be better conceived as addressing the philosophical implication of neuroaesthetics as it is currently conceived. The result of such an investigation will not only shed light on the intersection of science and culture but will also contribute to the debate about what form of naturalism is adequate for encapsulating the human world in all its enchanting variety.

Non-reductive Columbia naturalism

It may seem on first encounter with the term "naturalism" that it gains its meaning only in opposition to another realm of being, the "supernatural," and that in the absence of this opposition "naturalism" can have no meaning. However, it was the purpose of the naturalists working in America in the middle of the last century to reconstruct the meaning of naturalism without defining it in explicitly dualistic terms. Thus the rejection of supernaturalism by these thinkers cannot plausibly be seen as the rejection of one side of a dichotomy but rather as the dissolution of a supposed dualism on methodological principles. This understanding is perfectly in line with the perhaps only slightly exaggerated understanding of Dewey as "the destroyer of dualisms." One primary goal of the writers working to develop the "new naturalism" was to show that it did not lose all meaning in the absence of the juxtaposition with the supernatural or transcendent realm of being, but rather that this dualism had prevented understanding naturalism altogether. Therefore, the first step in understanding American naturalism is seeing it as opposed to all dualisms—man and nature, natural and supernatural, experience and nature. It is the rejection of this opposition that allows for the positive claim that anything that can be encountered by humans is accordingly "natural" and subject to investigation by the empirical (scientific) method. The attitude of seeing the natural as one side of an opposing dualism is the position that many naturalists rightly saw as preventing the extension of empirical methods of investigation to all areas of human discourse. Dewey summarizes this opposition in the first pages of the second edition of *Experience and Nature* as being the cause of the difficulty of some philosophers understanding his bringing together "experience" and "nature" into one comprehensive naturalistic metaphysics.

> To many the associating of the two words will seem like talking of a round square, so engrained is the notion of the separation of man and experience from nature. Experience, they say, is important for those beings who have it, but is too casual and sporadic in its occurrence to carry with it any important implications regarding the nature of Nature. Nature, on the other hand, is said to be complete apart from experience. Indeed, according to some thinkers the case

> is even in worse plight: Experience to them is not only something extraneous which is occasionally superimposed upon nature, but it forms a veil or screen which shuts us off from nature unless it can be "transcended." So something non-natural by way of reason or intuition is introduced, something supra-empirical. According to an opposite school experience fares as badly, nature being thought to signify something wholly material and mechanistic; to frame a theory of experience in naturalistic terms, is accordingly, to degrade and deny the noble and ideal values that characterize experience.[3]

For Dewey, the dualism of the natural and the supernatural, or more precisely of nature and experience, results in either having to transcend experience in order to access nature as it is in itself by way of some supra-empirical methodology, or even worse, to degrade the noble character of human experience by incorporating it within the purely material and mechanistic nature. These views can only be consistently maintained by supposing an intractable dualism between two realms of being where each asserts their own special domains and methods. However, if this dualism is rejected, then one is free to work out the implications of a naturalistic metaphysics without being committed to a devastating category mistake. Seen in this way, the dualism of the natural and the supernatural is an impediment to inquiry, and as such needs to be dissolved if any philosophical or scientific progress is to be made.

John Herman Randall, Jr., a student of Dewey and Frederick Woodbridge (another prominent figure in the history of naturalism in America) at Columbia, makes a similar point but further indicates the implications of this view for the application of scientific method. In the epilogue to the volume *Naturalism and the Human Spirit*, a book that came to be seen as the manifesto of the new naturalism, Randall writes:

> This double antithesis, between Nature and the Supernatural or Transcendental on the one hand, and between Nature and Man or human experience on the other, still governs present-day usage of "the natural" as a term of distinction. It is reinforced by our common employment of "natural science" to designate those methods and procedures which have been so successful in bringing knowledge of and power over what is not "supernatural," or "distinctively human." Conversely, those anxious for various reasons to assert the existence of a "Supernatural," or to insist on the unique and distinctive character of man or of some portion of his being—his soul or his mind—have been led to try to restrict the application of the methods of "natural science" to what is "merely natural," and above all to attempt to disparage their intrusion into human affairs. The unfortunate consequences in ignorance and failure in that field,

> springing from this divorce of two realms of being, Mr. Dewey has here once more vigorously set forth.[4]

Although Randall is here referencing Dewey's contribution to *Naturalism and the Human Spirit*, the point that the "ignorance and failure" in the attempt to restrict the intrusion of scientific method to human affairs is a familiar point in much of Dewey's writings. Further, the attempt to restrict the applicability of the scientific method to all of the various endeavors of human beings necessarily depends upon the separation of the human as something distinct from the nature. It is evident then that a primary point in understanding the naturalism developed by Dewey and the generation of younger philosophers that he influenced is the "refusal to take 'nature' or 'the natural' as a term of distinction."[5] This however, as Randall points out, is a purely negative definition of naturalism. On this view naturalism rejects any dualism between Nature and anything else and represents a categorical philosophical monism.

However, this seems to leave little space to clarify what nature consists in except to say that it is whatever human beings encounter in whatever various ways it is encountered. If this alone were the understanding of the meaning of naturalism, it would seem to lack the ability to discriminate between the natural and the supernatural at all and would merely be the theoretical representation of a philosophical prejudice against certain kinds of encounters. That is to say, it seems that the negative understanding of naturalism as the rejection of the dualism between the "merely natural" and the "supernatural" or "transcendent" makes the category of the "natural" so vacuous and all inclusive that there remains no grounds on which to reject religious revelation or rational intuition as a specific kind of encounter with "nature." If the only test for subsuming a phenomenom under the category of "nature" is that it is something which human beings encounter, then the rejection of naturalism as one side of a dichotomy with the supernatural only achieves folding those encounters which traditionally held supernatural warrant into the natural order of things. The naturalists would primarily respond that this critique implicitly rests on accepting "nature" as one side of a dualism and is not valid once that dualism has been rejected. Perhaps more satisfying however, this is exactly the kind of response that the negative understanding of naturalism seeks to elicit. The negative conception of naturalism does not deny that there are events that human beings experience that have traditionally been understood to possess supernatural warrant. It only denies that these kinds of events have reference to another order of being which stands above and beyond the "natural" order.

Once what was previously treated as supernatural or transcendent has been made part of the natural order by rejecting the dualism between the two realms, any previous experiences or events which possessed supernatural warrant under the protection of that dualism become subject to verification and inspection on the basis of the model that has proved successful in the sciences. Since the burden of scientific evidence requires that results can be repeated and independently verified, it makes claims that refer to another realm of being suspect on methodological grounds. Therefore, the rejection of the dualism between the natural and the supernatural does not make "nature" a vacuous category but rather provides the means through which to subject all claims to knowledge to the standards of evidence and inspection that have proven themselves effectual in the sciences.

If naturalism were only a negative thesis, it would hold little interest as a philosophical theory. However, there is also a positive sense of naturalism that weds it inextricably to the scientific method. If "nature" is said to consist in all the various things that factor into human experience, then it is the role of naturalism to seek an understanding of the nature and relationship between the various aspects that comprise nature. This view might be classified as Dewey's "metaphysical anthropocentrism." Randall states this positive aspect of naturalism as follows:

> Naturalism thus merges in the generic of philosophy as critical interpretation—the examination of the status of all these varieties of "stuff" in Nature—or Being, or in Reality—and the discovery of their various relations to each other and their respective functions in man's experience. Positively, naturalism can be defined as the continuity of analysis—as the application of what all the contributors call "scientific methods" to the critical interpretation and analysis of every field.[6]

Naturalism as continuity of analysis represents the positive thread of the naturalism developed in the mid-twentieth century. It should be clear how it is closely related with the opposition to supernaturalism, as it is only the things that figure as part of human experience that can be subjected to inquiry. Randall is also referencing Dewey's view that philosophy is the "criticism of criticism." It is only when philosophy is wedded to the methods of science that it can carry out this critical task.

However, the use of the phrase "scientific method" by the naturalists is sufficiently vague as to make it unclear what their conception of scientific method exactly is. At the very minimum, the use of the phrase means that empirical evidence should be brought to bear in all areas of inquiry. Thelma Lavine states that the methodological principle of naturalism uses "methods which administer the checks of intelligent experiential verification in accordance with the contemporary criteria of objectivity. The significance of this principle does not lie in the advocation of empirical method, but in the conception of the regions where that method is to be employed."[7]

Since the naturalists saw their novel contribution not as advocating adopting the scientific method (which had already proven itself in reference to understanding physical reality) but rather for its extension, it is perhaps not surprising that they are less than careful in enumerating what is exactly entailed by "intelligent experiential verification." It is also important to see that many of the writers who contributed to *Naturalism and the Human Spirit* were very much under the strong influence of Dewey and presupposed many of Dewey's views in their writings. Thus the methodological principle of the continuity of analysis, which is asserted by the contributors to the volume, is informed by Dewey's views on method, to which I shall turn shortly. However, before turning to an analysis of Dewey's view on method and what it exactly means to say that it should be extended to all areas of inquiry, it will be useful to return to Randall once more to provide a summary of the attitude of the new naturalists toward science and their core belief that its methods can be expanded to deal with the difficulties of human culture.

> Today we are at last in possession of a science that insists on the importance and reality of all man's experience and enterprises, and has developed concepts that promise to render them all intelligible. In consequence we are now able to erect for ourselves philosophies that can find a natural and intelligible place for all human interests and aims, and can embrace in one natural world, amenable to a single intellectual method, all the realities to which human experience points: symphonies as well as atoms, personality as well as reflex action, religious consecration as well as the laws of motion or the equations of field theory. These contemporary naturalistic philosophies are thus a direct function of present maturing of the scientific enterprise.[8]

It was not philosophic prejudice against the supernatural or transcendent that motivated the new naturalism in America. Rather, it was a direct result of the growth of science itself. Once science was capable of incorporating all the rich diversity of human experience through a "single intellectual method," philosophy, as critical interpretation, became able to address the place and relations of those experiences both with each other and with the aims and purposes of human living itself.

Dewey's discussions on method are sprinkled throughout much of his corpus, at times making it difficult to glean a comprehensive understanding of his views on the subject. This does not represent organizational carelessness on Dewey's part but rather indicates the overarching importance of empirical (experimental) methodology to his entire philosophical endeavor. Dewey's writings on ethics, aesthetics, democracy, education, and metaphysics are best viewed as applications of his views on methodology rather than as direct explications of his methodological position. However, investigating Dewey's writings in two areas, the role of method in his naturalistic metaphysics and his understanding of logic as the general pattern of inquiry, will illustrate the general commitments that Dewey held about the scientific method and its applications. In particular, examining Dewey's notion of the *situation* and the process of a situation going from indeterminate, to problematic, to determinate will help to make clear his position on method, in particular the role that context plays in his formulation of inquiry.

An important theme in Dewey's discussion of method, both in *Experience and Nature* and *Logic: A Theory of Inquiry*, is that of continuity. In *Experience and Nature*, especially the first chapter, Dewey focuses on the continuity between scientific method and common sense. In the *Logic* however, Dewey defines continuity in a more fundamental sense—that of the continuity between culture and nature. In fact, by the time of his death, Dewey was contemplating the use of "culture" to refer to both. Since the logic of scientific discovery relies on this more basic meaning of continuity, it will be instructive to note how Dewey defines the term. Further, although Dewey is specifically referencing a naturalistic theory of logic the point applies to Dewey's naturalism generally. Continuity, then, is a central tenet of Dewey's naturalism.

> The primary postulate of a naturalistic theory of logic is continuity of the lower (less complex) and the higher (more complex) activities and forms. The idea of continuity is not self-explanatory. But its meaning excludes complete rupture on the one side and mere repetition of identities on the other; it precludes reduction of the "higher" to the "lower" just as it precludes complete breaks and gaps. The growth and development of any living organism from seed to maturity illustrates the meaning of continuity.[9]

Dewey's use of a biological example to illustrate the meaning of continuity is important because it serves as a framework in which to understand how logic is continuous with biological functioning and how scientific method is continuous with common sense. The mature summer tomato plant is not identical to the seedling from which it grew, but neither is it separated by an ontological gap; rather it is a process. In the same sense, logical inquiry is continuous with the biological interaction of an organism with its environment in that logic presupposes a biological basis. This is so because for Dewey an organism interacting with its environment requires judgment, if even in a proto-logical sense. Yet, although the higher forms are continuous with the lower, in that the former emerges from the latter, they are not identical nor can the one be reduced to the other. Higher forms preclude explanation exclusively in reference to the lower just because they are continuous.

The principle of continuity is essential in understanding Dewey's non-reductive naturalism in that it precludes treating higher-order phenomena, such as aesthetic experience, as being identical with biological structures from which they emerge. This point will be important in assessing the project of neuroaesthetics from a Deweyan perspective. Whereas ontological naturalism focuses on the reduction of entities or events characteristic of "higher" forms to what is supposed to be a more fundamental level of explanation, Dewey's understanding of continuity clues us in to the idea that the emergence of higher-order functioning is continuous with lower. The two are connected in the same way that youth is connected to adulthood; they are both part of a single process that unfolds in time and hence a complete understanding of the whole resists reducing one to the other. While the language of reductionism lends itself to speaking of mental events being nothing over and above physical events, Dewey's process metaphysics cannot be adequately understood in these terms.

While Dewey specifically focuses on the continuity of the biological and the cultural, the same point can be applied to the relationship between science and everyday experience. Although science is an abstraction, for Dewey it is always grounded in, and responsible to, the existential matrix of lived experience. Rather than focusing on the results of scientific inquiry, that is, abstract scientific objects, in elucidating method, Dewey focuses on the actual process of inquiry as carried out by scientists. By focusing on the way in which scientific methodology is continuous with experience in a more general sense, Dewey is able to show how the examination of scientific inquiry illustrates the goings-on in any experience. Sandra Rosenthal nicely captures this essential aspect of Dewey's thinking on method as follows:

> All experience is experimental, then, not in the sense that it is guided by sophisticated levels of thought, but in the sense that the very structure of human behavior both as a way of knowing and as a way of being embodies the features revealed in the examination of scientific method. It is not that human

> experience, in any of its facets, is a lower or inferior form of scientific endeavor, but rather that scientific endeavor, as experimental inquiry, is a more explicit embodiment of the dynamics operative at all levels of experience and hence the ingredients are easier to distinguish. The pursuit of scientific knowledge is an endeavor throughout which are "writ large" the essential characters of any knowing, and it partakes of the character of the most primal modes of activity by which humans participate in creatively structuring their world. An examination of scientific method, then, provides the features for understanding the very possibility of its existence as emerging from rudimentary experience.[10]

Studying the structure of scientific methodology does not produce the problem of reconciling the abstract, intellectualized apparatus of science with the more rudimentary forms of experience found in everyday experience as it is lived. Rather, it exposes the very possibility of the emergence of scientific method from experience. The claim that all experience is experimental is not to claim that the mode of everyday experience is scientific but only that scientific practice is continuous with less sophisticated human undertakings. Scientific method becomes applicable to all areas of inquiry only when it is seen in this way. Only when science is taken to be an illumination of the very structure of experience at all levels is the tension between science and common sense dissolved. In fact, for Dewey, there could be no such tension because science emerges from the same patterns of living that constitute all types of experience. Thus, by bringing scientific method to bear on all aspects of experience, when experience is already conceived as experimental in nature, methodological naturalism can refine less abstract modes of experience and thereby bring them under the control of intelligence. Dewey asserts the continuity between common sense and scientific method as follows:

> Scientific subject-matter and procedures grow out of the direct problems and methods of common sense, of practical uses and enjoyments, and react into the latter in a way that enormously refines, expands, and liberates the contents and the agencies at the disposal of common sense. The separation and opposition of scientific subject-matter to that of common sense, when it is taken to be final, generates those controversial problems of epistemology and metaphysics that still dog the course of philosophy. When scientific subject-matter is seen to bear genetic and functional relation to the subject-matter of common sense, these problems disappear.[11]

It is important to note that Dewey's conception of scientific method and its continuity with common sense relies on the rejection of the dualisms of theory/

practice and mind (knower)/world (what is known). Since experience as experimental is, for Dewey, transactional, more specifically a transaction between an organism and its environment, all human activities are instrumentalities, directed toward manipulating and structuring the environment in ways that are amendable to human ends. Seen as such, even the laboratory procedures of the scientist are shot through with values that are directed toward the understanding, refinement, and control of problems that emerge from common sense. Dewey's methodological naturalism is related to his larger project of the reconstruction of philosophy and the dissolution (rather than resolution) of traditional philosophical problems that are predicated on the bifurcation of the realm of activity from the realm of reflection.

Dewey's methodological naturalism, understood as the claim that the procedures of science can be fruitfully extended to all areas of inquiry, *requires* the continuity between the *subject-matter* of science and common sense. To claim otherwise would be to reinforce the metaphysical break between doing and knowing, which Dewey rightfully identifies as the source of the separation of man from nature and fact from value. All human activity is value-laden because of the very transactional nature of experience itself. There is no value-free objective world that we can simply read, free from the specific contexts and purposes of human activity.

Part of what provides for the continuity of scientific inquiry and common sense is that for Dewey all inquiry has the same pattern. Thus it is essential to indicate in outline the pattern of inquiry that Dewey thinks is common to all thinking. Dewey most clearly elaborates the pattern of inquiry in the *Logic* although it is also developed in the earlier publications, *Studies in Logical Theory* and *How We Think*. What is of interest here in examining Dewey's theory of inquiry is how it firmly locates him within the sphere of *methodological naturalism*. Before discussing Dewey's pattern of inquiry it is important to reemphasize a different sense of Dewey's use of the term continuity—the continuity of the physical or biological environment with the cultural environment. After that discussion, it will be possible to fully explicate Dewey's notion of *the situation* and how it relates to the pattern of inquiry.

Although I have thus far been emphasizing the continuity between common sense and scientific inquiry in Dewey's thought, there is a more central sense of continuity—that between the biological and the cultural. A central tenet of Dewey's naturalism is that higher-order forms, whether they are logical or even aesthetic forms, are continuous with the lower level biological functioning of an organism with its environment. The emergence of logical and aesthetic experience is closely associated with intelligence. It is when an organism intelligently interacts with its environment that the merely biological can become the cultural. This is one reason why Dewey's naturalism is non-reductive.

Although logic and aesthetics are not separated from either everyday experience or biological interaction by an ontological gulf, it does not follow that they can be understood using biological or physical explanations alone. This is because, for Dewey, the environment in which the organism lives is not only biological but also cultural. So while Dewey rightly stresses that logic or method does not intrude upon experience from without, that it is continuous with and a product of biological activity, because the environment is suffused with cultural meanings and objects, a naturalistic account of experience cannot be properly understood without reference to culture. Dewey makes this point in the *Logic* in the following way.

> The environment in which human beings live, act, and inquire, is not simply physical. It is cultural as well. Problems which induce inquiry grow out of the relations of fellow beings to one another, and the organs for dealing with these relations are not only the eye and ear, but the meanings which have developed in the course of living, together with the ways of forming and transmitting culture with all its constituents of tools, arts, institutions, traditions, and customary beliefs.[12]

When Dewey claims that inquiry is the paradigmatic way that intelligent organisms interact with their environment, it is essential to understand his use of "environment" in this expansive sense. Without the ability to use and employ cultural meanings, intelligence would not be possible. In fact, Dewey goes as far as to claim that the changes wrought by the cultural environment on a biological organism transform merely organic behavior into intelligence. Yet these changes occur *within* the organism. When philosophers reflect on the phenomena of understanding speech, perceiving the expressive qualities of a work of art, or engaging in politics, it becomes clear that human beings are a hybrid of biology and culture. It is because the environment in which we live is both physical *and* cultural that intelligence emerges. "This modification of organic behavior in and by the cultural environment accounts for, or rather is, the transformation of purely organic behavior into behavior marked by intellectual properties."[13] While Dewey does not deny that purely organic behavior takes place, for instance, withdrawing one's hand from a flame, or sneezing in the presence of an allergen, even this behavior is suffused with cultural meaning.

The alteration of organic behavior always takes place in a physical and cultural environment. When this modification is also effected *by* the cultural environment, meaning which transcends the merely physical environment comes forth. Dewey, however, is not blind to the criticism that the large differences between the behavior and activities of human beings and that of other animals cannot be accounted for simply by positing that human beings live in a cultural as well as a physical environment, for this alone cannot explain how the cultural environment comes to be in the first place. However, this difference is not cause to posit a supernatural source, which is discontinuous with experience, as the explanation for the distinctive powers of human beings. Rather Dewey's problem is explaining the development of human capabilities out of simpler biological or organic behaviors. In order to do so, Dewey focuses on the development of language as the key to understanding the continuity between the biological and the cultural.

For Dewey, language means more than oral and written communication. If language is to be the key to understanding the transformation of the merely biological environment into the cultural, and also for how meaning is created and transmitted in general, then it must have a broader specification than just speech. This point is not only significant for Dewey's theory of inquiry but for his aesthetic theory as well. Language, of course, includes speech,

> but it includes also not only gestures but rites, ceremonies, monuments, and the products of industrial and fine arts. A tool or machine, for example, is not simply a simple or complex physical object having its own physical properties and effects, but it is also a mode of language. For it *says* something, to those who understand it, about operations of use and their consequences.[14]

Dewey's conception of language is *operational* or *functional*. At minimum, Dewey claims that language operates as the mode of communication in that it is employed in the making of something common. Understanding language in this broadest sense requires that one takes on the standpoint of the other. Whether the medium of communication is a sound or a work of art, in order to understand what it says requires that one understands how it functions in experience.

Since the context in which communication takes places is the cultural environment, when language operates in experience it does so not merely as the physical thing that it is but also as a bearer of meaning. This point is central in understanding the non-reductive cultural ontology that is at the heart of Dewey's naturalism. Dewey makes this point as follows: "Language is made up of physical existences; sounds, or marks on paper, or a temple, statue, or loom.

But these do not *operate* or function as mere physical things when they are media of communication. They operate in virtue of their *representative* capacity or *meaning*."[15] In other words, language functions as culture, not as the physical things in which culture is embodied and transmitted. Dewey's point is that, in possessing meaning, a physical object, whether it is a sound, a line, or even food, functions as a cultural object. The relationship between the two is the reciprocal relation of embodiment and emergence. The meaning is embodied in a physical thing and emerges from it but is not identical to this underlying physical medium because of the way that it functions. If the environment in which we live was exclusively physical or biological, as Dewey denies, then these objects would not function as media of communication. Since, however, the environment with which the human organism interacts is also cultural, these physical things are transformed, via their representational function, into meanings embodied in a physical medium. It is through communal, functional use that things get their meaning, that nature is transformed into culture.

Having explicated Dewey's view regarding the continuity between scientific inquiry and common sense, and between the physical and cultural environment, we are now in a position to make sense of his often maligned and misunderstood use of *the situation* as a technical term. According to Dewey, "what is designated by the word 'situation' is *not* a single object or event or set of objects and events. For we never experience nor form judgments about objects or events in isolation but only in connection with a contextual whole. This latter is what is called a 'situation.'"[16] Dewey's special use of the word *situation* indicates that his naturalism is thoroughly contextual. Since objects and events can only be understood as connected with a contextual whole, it is impossible to inquire about any object outside of the cultural environment. Since the cultural environment, as has been shown, is distinguished from the merely physical environment through the former possessing meaning—that is, being representational or symbolic—any method of inquiry that focuses on the object to the exclusion of the context will generate untenable dualisms. In experience as it is actually lived, there is never found an isolated object. An object or event may stand out from the contextual whole as being particularly problematic and in need of resolution, but the end-in-view of this resolution always refers back to the contextual whole of experience rather than the object abstracted from the environment in which it is embedded.

The situation, for Dewey, is always marked by a qualitative wholeness (a claim that will be emphasized in a different context in Dewey's account of aesthetic experience) and objects or events within this contextual whole are isolated (in a methodological not a metaphysical sense) only insofar as they may serve as clues for the way in which a situation can be used or enjoyed. It is Dewey's emphasis on aesthetic experience as the key indication of what it means for a situation to exhibit a qualitative wholeness that can be used to individuate situations, that makes understanding his aesthetics so central to the broader understanding of his naturalistic account of experience.

However, understanding aesthetic experience does not *exhaust* Dewey's meaning of "the situation," it rather exemplifies it. Dewey's meaning of the situation is more general than his account of aesthetic experience because it applies to situations that are themselves not considered aesthetic. Yet both make use of the idea of qualitative wholeness of *an* experience to mark it off from other situations as a subject of inquiry. Dewey's notion of the qualitative wholeness of an aesthetic experience is crucial to understanding his naturalism. Since this non-reductive version of naturalism, which recognizes the continuity between the biological and cultural realms, provides a compelling framework in which to evaluate the project of neuroaesthetics, it will be important to understand the relationship between science and art in Dewey's thinking before turning to a direct discussion of neuroaesthetics.

The relationship between art and science in Dewey's pragmatic naturalism

In the chapter "Experience, Nature and Art," in Dewey's masterwork, *Experience and Nature*, Dewey goes a long way in clarifying his understanding of the relationship between science and art. Beginning with his usual criticism of Greek philosophy for equating knowledge with contemplation instead of practice or art, Dewey goes on to draw the conclusion that, from the vantage point of Greek philosophy, art is subordinate to science. If however, this hierarchy is inverted and the creative powers operative in art are seen as primary to the contemplation of final and unchanging ends "it would then be seen that science is an art, that art is practice."[17] Dewey continues that then:

> the only distinction worth drawing is not between practice and theory, but between those modes of practice that are not intelligent, not inherently and immediately enjoyable, and those which are full of enjoyed meanings. When this perception dawns, it will be a commonplace that art—the mode of activity that is charged with meanings capable of immediately enjoyed possession—is

> the complete culmination of nature, and that "science" is properly a handmaiden that conducts natural events to this happy issue. Thus would disappear the separations that trouble present thinking: division of everything into nature *and* experience, of experience into practice *and* theory, art *and* science, of art in useful *and* fine, menial *and* free.[18]

This passage is incredibly rich with several central Deweyan themes. Not only does Dewey recognize the Greek tendency to disparage practice as the origin of a series of dualisms that have, to his mind, plagued philosophy ever since, he goes further to make the startling claim that science ought to be properly understood as instrumentally valuable insofar as it helps nature bring forth meanings that are immediately enjoyable. In fact, Dewey claims that a practice is *intelligent* insofar as its meanings can be immediately enjoyed. Thus intelligence for Dewey is, at its core, *aesthetic*. Since it is the primary characteristic of art and the aesthetic that aesthetic meanings are pervasively *felt* meanings, Dewey is claiming that scientific inquiry is successful when its conclusions are not seen as isolated from the course of ordinary experience but instead culminate in the sort of felt enjoyment that is more closely associated with the experience of art.

The recognition that scientific inquiry, as a mode of experience, is successful when it mirrors the rhythms that constitute the aesthetic is evidence of, as John Onians describes, Dewey's "anticipation that one day even the latest discoveries of the rhythms of physics may become the subject of poetry and implicitly of art."[19] While it is true that Dewey recognizes that there is no special subject-matter of the arts and that any experience, even scientific, is open to become the subject of the arts, Onians misses the deeper philosophical subversion of Dewey's inversion of art and science. Neither art nor science is cut off from the rhythms of exchange between an organism and its environment that constitutes experience. It is this basic rootedness of art and science in the biological matrix of organism and environment that allows Dewey to claim that intelligence is at root aesthetic. This is because for Dewey, it is aesthetic experience, in its rarified form, that illustrates the basic properties of experience in the raw.

> Esthetic experience is experience in its integrity. Had not the term "pure" been so often abused in philosophic literature, had it not been so often employed to suggest that there is something alloyed, impure, in the very nature of experience, we might say that esthetic experience is pure experience. For it is experience freed from the forces that impede and confuse its development as experience; freed, that is, from factors that subordinate an experience as it is directly had to something beyond itself. To esthetic experience, then, the philosopher must go to understand what experience is.

> For this reason, while the theory of esthetics put forth by a philosopher is incidentally a test of the capacity of its author to have the experience that is that subject-matter of his analysis, it is also much more than that. It is a test of the capacity of the system he puts forth to grasp the nature of experience itself.[20]

Since science is itself a form of practice, of experience, and it is through an understanding of aesthetic experience that we may come to an analysis of the nature of experience itself, it follows that it is through the arts, as Dewey claims, that we will best come to understand the nature of scientific practice. This claim, as shall be seen in the following section, will have significant implications for a Deweyan analysis of the problems and prospects of neuroaesthetics.

The problems and prospects of neuroaesthetics

Martin Skov and Oshin Vartanian, in an introduction to one of the most comprehensive anthologies currently available on neuroaesthetics, define it as "a certain way of doing aesthetics, using neuroscience as a method of inquiry where other approaches to aesthetics have used philosophical analysis or strictly psychological models. As such, neuroaesthetics studies how aesthetics behavior is underpinned by brain process."[21] What is important to highlight about this definition is the way in which Skov and Vartanian conceive of neuroaesthetics as methodologically opposed to philosophical aesthetics rather than continuous with it. If, as this formulation of the definition of neuroaesthetics suggests, the field is conceived of as a replacement of, rather than a supplement to, traditional approaches to the study of the arts and aesthetic experience, there is a legitimate concern that neuroaesthetics is conceived by its practitioners as a reductionist science.

A further question of debate in the early formulations of neuroaesthetics is whether or not an art-based approach is preferable to one that has a broader understanding of the range of aesthetic objects that are of interest to human beings. Steven Brown and Ellen Dissanayake, claim that neuroaesthetics, at the current stage of its development seems to be "expressly concerned with art, not least the masterpieces of European visual art and even abstract paintings" rather than the broader notion of aesthetic experience itself.[22] They go on to claim that neuroaesthetics, as presently conceived, is too narrowly focused on art objects and that this represents the influence and bias of Enlightenment theories of the arts, which have thus far precluded the study of non-western arts that are not clearly associated with the disinterested contemplation of beauty. This has led not only to a focus on art objects in many influential early approaches to neuroaesthetics but also to an emphasis on the study of beauty.[23] As such, neuroaesthetics has often been conceived of as the search for the discrete neural basis of a universal theory of beauty. This approach to neuroaesthetics, called *parallelism*, is often associated with the work of Semir Zeki.

In summarizing recent work in neuroaesthetics, Anjan Chatterjee provides a concise statement of the view. "The parallelism approach to neuroaesthetics recognizes that the production and perception of art ought to conform to principles of neural organization. Properties of artworks and strategies used by artists have parallels in how the nervous system apprehends and organizes its visual world."[24] In attempting to derive an answer to the question of why human beings possess the capacity for vision, Zeki himself introduces the central thesis of parallelism. "It is the answer to that question that immediately reveals a parallel between the functions of art and the functions of the brain, indeed ineluctably drives us to another conclusion, that the overall function of art is an extension of the function of the brain."[25] Zeki claims that the overarching function of the brain is to acquire knowledge of the world and that in order to do so the brain must select from ever changing perceptual information to arrive at a constant understanding of an object. This principle of constancy of the brain in turns allows us to categorize objects so that we may recognize other objects that belong to the same kind or class—even though the visual information that reaches the brain in each case is in a state of constant change.

Given the thesis of parallelism, understanding the function of the brain in turn provides a definition of the function of art.

> The definition that I have given above of the function of the visual brain—a search for constancies with the aim of obtaining knowledge about the world—is applicable with equal vigor to the function of art. I shall thus define the general function of art as a search for the constant, lasting, essential, and enduring features of objects, surfaces, faces, situations, and so on, which allows us not only to acquire knowledge about the particular object, or face, or condition represented on the canvas but to generalize, based on that, about many other objects, and thus acquire knowledge about a wide category of objects or faces.[26]

It should not come as a surprise that a scientist writing about art would express an aesthetic cognitivism as the function of art. Further, Zeki's parallelism heavily informs his preference for philosophers of art, notably Plato, Hegel, Kant, and Schopenhauer. Each of these thinkers expresses some sort of Absolutism or

Universalism about the purpose of art consisting in reaching beyond the world of appearances and particulars to the more permanently stable reality that lies beyond all such appearance. The difference being that instead of searching for the principle of constancy in some metaphysically real Platonic Ideal or Hegelian absolute, Zeki locates this in the functioning of the brain.

The problem with parallelism lies in assessing the explanatory power that the functional parallels between artworks and the brain possess. If, as Zeki claims, artists are neurologists studying the functioning of the brain by utilizing artistic methods, then the study of their works would be revelatory about the brain but not necessarily about art. This poses a central question in understanding the prospects of neuroaesthetics. Is neuroaesthetics essentially concerned with the task of understanding the brain by studying how the brain processes art or is neuroaesthetics meant to tell us something about art and the aesthetic? The direction of explanation is a crucial question in assessing parallelism as an approach to neuroaesthetics. This problem is compounded by the emphasis on beauty in neuroaesthetics. Artworks are produced for many functions, certainly chief among them provoking an experience of beauty and acquiring knowledge about the world, especially knowledge that might be better gleaned through visual, audible, or somatic experiences. However, equating the investigation of the neural underpinnings of our experience of art exclusively with one or both of these functions problematically narrows the scope of what human beings find valuable in encounters with art.

This point can be illustrated by looking at Zeki's use of the mobiles of Alexander Calder to exemplify parallelism. The processing of motion by the brain is centered in Area V5 of the visual cortex. Zeki claims that Calder's de-emphasis of color in the construction of his mobiles, focusing primarily on the use of white and black, might serve as evidence for his thesis that artists utilize the methods of their craft to learn something about the brain. Zeki writes that:

> kinetic art, especially in the hands of Calder, resulted in works that act as a perfect stimuli for the cells of V5. Another important feature that perhaps reinforces the view that I present here—that artists try to learn something about the organization of the visual brain, through the techniques unique to them—is found in the general emphasis on movement and in a de-emphasis of color and form, mirroring so well the physiology of V5.[27]

The claim that Calder's work serves as a perfect example to illustrate the functioning of Area V5 is a point that illuminates the way that the brain processes motion and is certainly of interest to neuroscientists interested in understanding the visual brain.

However, the second point that Zeki raises, that it was Calder's intent to learn something about the visual brain because of parallels that exist between the sparse use of color in his mobiles and the functioning of V5 as an area that does not pay attention to color, is significantly less interesting as a thesis about aesthetics. Bevil Conway and Alexander Rehding make a similar point, drawing out the consequences in terms of beauty. "An Alexander Calder sculpture may consist of optimal stimuli for the brain's motion center, but this aspect of the work does not make it beautiful. The art simply provides a fascinating demonstration of the computations of the brain's motion-perception circuits, and the genius of the artists for discovering them."[28]

Artworks may provide interesting examples of specific function of the brain—motion, color, and facial recognition—of genuine interest to neuroscientists and art theorists alike, but this does not equate to providing a universal foundation for judgments of beauty, the goal of Kantian aesthetics, on this basis alone. As we have seen, parallelism in neuroaesthetics is the thesis that the function of art is parallel to the function of the brain and that in turn artists are engaged in investigating the brain through the techniques and methods of art making. Parallelism in turn rests on a form of *aesthetic* functionalism, the view that the function of works of art is to produce the experience of beauty. Taken conjointly, these views suggest that one—if not the primary—task of neuroaesthetics is to discover the neural basis for universal judgments of beauty. As we shall see shortly, this project runs the risk of scientific reductionism.

Hideaki Kawabata and Semir Zeki performed one such study that investigated the neural underpinnings of judgments of beauty using fMRI techniques. In their introduction to the article "Neural Correlates of Beauty," they explicitly invoke the Kantian project in aesthetics. "This work is an attempt to address the Kantian question experimentally by inquiring into whether there are specific neural conditions implied by the phenomenon of beauty and whether these are enabled by one or more brain structures."[29] In order to avoid questions of how the experience of beauty might be culturally informed, Kawabata and Zeki had subjects rank 300 paintings as beautiful, ugly, or neutral. These samples included examples of abstract paintings, landscapes, still life, and portraiture. Subjects were then shown 192 samples while being monitored by fMRI and the results were correlated with their previous reports of beauty and ugliness. Their results showed that "the judgment of a painting as beautiful or not correlates with specific brain structures, principally the orbito-frontal cortex, known to be engaged during the perception of rewarding stimuli." Further, their results showed that judgments of ugliness do not activate a different part of the brain than judgments of beauty but "that it is rather a change in relative activity in the orbito-frontal cortex that correlates with the judgment of beauty and ugliness."[30]

This result is of particular interest in addressing whether or not there can be a positive judgment of ugliness or if rather a judgment of ugliness is merely the negation of a judgment of beauty. That the same area of the brain is engaged in judgments of beauty and ugliness suggests that aesthetic judgments take place on a continuum that correlates to the relative amount of activity in the orbito-frontal cortex. Kawabata and Zeki are careful to point out the limitations of an fMRI study in that it does not detect other areas that may contribute to performing the task of aesthetic judgment. Nevertheless, Conway and Rehding point to the reductionist worry in using fMRI techniques to establish the areas of the brain that are responsible for judgments of beauty.

> Rational reductionist approaches to the neural basis for beauty run a … risk of pushing the round block of beauty into the square hole of science and may well distill out the very thing one wants to understand. There is a popular conception of beauty as a fixed attribute of objects, a notion that much of current neuroaesthetics depends upon. But there is a distinction between abstract notions of beauty and our experience of it … Beauty is an analog, not binary, condition that varies in complex ways with exposure, context, attention, and rest—as do most perceptual responses. In trying to crack the subjective beauty nut with scientific, objective information, we also run the risk of fueling a normative, possibly dangerous campaign through which science is required to valorize our experience. Should we deny someone's experience of beauty if the mOFC (medial orbito-frontal cortex) is not activated? Obviously not.[31]

If as Dewey suggests an aesthetic experience is the result of the interaction between an organism and its environment that results in a consummatory experience, then it is hard to see how being flashed images of paintings while being monitored by a fMRI machine could capture the contextual (including environmental) features necessary for an aesthetic experience. In fact, the necessities of objective laboratory studies almost require that any experience that the subject has will be inchoate rather than consummatory. It remains to be seen if neuroaesthetics will be able to devise experimental techniques that can capture the cultural context of aesthetic experience. But, as I argued in the previous section, following Dewey, one cannot abstract an aesthetic experience from the cultural context in which it takes place. If this is so, then inducing an experience of beauty in the laboratory in order to correlate brain states to aesthetic judgments raises the question if what is being measured is aesthetic at all. Thus it seems only prudent to resist reducing the experience of beauty to the activation of the mOFC, while simultaneously recognizing that part of the biological activity that comprises an aesthetic experience as a whole may consist of such activity.

Further, there are methodological concerns surrounding correlating brain activity to the subjective reports of experimental subjects as to whether or not a particular painting is beautiful or ugly. By showing subjects paintings, which are paradigmatically thought of as aesthetic objects, it is unclear if the results of experiments like those performed by Kawabata and Zeki are measuring aesthetic responses in a non-question begging way. This concern is likewise a result of the art-centric approach of much of neuroaesthetics.

This art object-based approach is clearly at odds with Dewey's insistence that art is not an object but an activity or experience that is set apart from the flow of ordinary experience by a qualitatively felt unity that marks it off as *an experience*. If one task of neuroaesthetics is to determine which philosophical theory of aesthetic experience is more consistent with what goes on in the brain, then the object-centric bias of neuroaesthetics unwittingly sways results toward certain theories of art. As Brown and Dissanayake put the point, "as presently conceived, neuroaesthetics has no way of distinguishing art from nonart. Characterizing the neural responses elicited by viewing a modern abstract painting begs the question that what is being assessed is a response to art."[32]

If neuroaesthetics is incapable of addressing one of the central questions of the philosophy of art, the distinction between art and non-art, in a non-question begging way, then it is clear that neuroaesthetics must be understood in a non-reductive way. Insofar as the parallelism approach to neuroaesthetics as presently conceived reduces the aesthetic to the experience of the art object, it is too narrowly construed and reliant on accounts of the arts derived from more traditional approaches to aesthetics, such as aesthetic functionalism and Kantianism. Since significant disagreement exists as to the adequacy of the Kantian project of grounding the conditions for universal yet subjective judgments of taste, and one variant of neuroaesthetics seems wedded to experimentally verifying this Kantian conceit, then neuroaesthetics is incapable of addressing, on its own terms, whether or not the experimental data gleaned from correlating brain states with a subject's response to works of art is in fact illuminating what goes on in the brain during an aesthetic experience.

Dewey himself recognizes this point when he warns against the compartmentalization of art from the normal process of living. Here is Dewey:

> The understanding of art and of its role in civilization is not furthered by setting out with eulogies of it nor by occupying ourselves exclusively at the outset with great works of art recognized as such … Theory can start with and from acknowledged works of art only when the esthetic is already compartmentalized, or only when works of art are set in a niche apart instead of being celebrations, recognized as such, of the things of ordinary experience. Even crude experience, if authentically an experience, is more fit to give a clue to the intrinsic nature of esthetic experience that is an object already set apart from any other mode of experience.[33]

Here Dewey and Dissanayake both share a concern with using art objects as the starting point for theorizing about aesthetic experience. To this extent, approaches to neuroaesthetics that begin with an investigation of the biological processes involved in experiencing a work of art conceived as such, is an instance of the type of compartmentalization of art from the ordinary course of experience that Dewey thinks will not lead to a fruitful understanding of art. This point is further complicated by the fact that it will be difficult to know if the experience that a subject reports as having is in fact an aesthetic experience. The presumption that, merely in virtue of using works of art as experimental material, we will therefore be drawing conclusions about aesthetic experience is, at this stage of development of neuroaesthetics, troubling.

It is important to emphasize that, in this chapter, I have only focused attention on one approach to neuroaesthetics. There is a significant and growing literature on dance, music, evolutionary approaches to the arts, and the effect of cognitive impairments on artistic activity, among other topics that make up the field of neuroaesthetics. Further, many active neuroaesthetic investigators reject the art-based approach to neuroaesthetics that I have been criticizing. For instance, in the introduction to a recent special issue on neuroaesthetics in *Psychology of Aesthetics, Creativity and the Arts*, which collects an impressive array of essays that exhibit the breadth of current approaches to neuroaesthetics, Marcos Nadal and Martin Skov reject the art-centric approach to neuroaesthetics.

> There is a fundamental reason why we believe that establishing the boundaries of neuroaesthetics along the art–not art division betrays the whole purpose of neuroaesthetics. The notions of art, artist, and artwork are culturally and historically contingent, as are the criteria used to distinguish what is an artwork from what is not, or what is an art form from what is not … Rather, we believe

> neuroaesthetics should focus on a certain kind of way objects can be experienced when people approach them with an "aesthetic attitude."[34]

While Nadal and Skov's more inclusive approach to defining the scope of neuroaesthetics is welcome, they nevertheless trade one contested approach to aesthetics for another. By invoking the idea of an aesthetic attitude, a view that traces its philosophical parentage to the formalism of Kant, and one that has been widely rejected in the wake of George Dickie's criticism, Nadal and Skov are trading one philosophical mistake for another.[35]

If the future of neuroaesthetics is to truly be interdisciplinary, then neuroscientists investigating the neural correlates of aesthetic phenomena should draw inspiration more widely from philosophical theories about art and the aesthetic. Without a grounding in the best approaches to the arts from the humanities, neuroaesthetics runs the risk of becoming a reductive science that, rather than contributing to our understanding of what makes aesthetic experience such an essential and paradigmatic aspect of the human experience, weakens our understanding of the complexity of our aesthetic lives by replacing the richness of the aesthetic dimension as it is qualitatively lived, for a version of aesthetics that tells us nothing about why human beings seek out aesthetic encounters at all.

Conclusion

In this chapter, I have focused on two critiques of parallelism in neuroaesthetics—the art-centric approach to aesthetic experience and the danger of reducing judgments of beauty to the activation of specific areas of the brain. Both criticisms emerge from the basic supposition that the function of art mirrors the function of the brain. By invoking a Deweyan account of scientific inquiry and aesthetic experience, and the non-reductive naturalism that underlies both of these views, it is possible to understand the project of neuroaesthetics in a way that avoids both of these problems.

If neuroaesthetics is to be a truly interdisciplinary research agenda into the nature and value of art and the aesthetic, then it must take account of the best that philosophical inquiry into these areas has to offer. I have argued that a Deweyan account of aesthetic experience, that stresses the emergence of the aesthetic from the interaction between an organism and its natural *and* cultural environment, provides a compelling framework from which to conceptualize the task of neuroaesthetics in a non-reductive manner.

Neuroaesthetics is a truly exciting approach to the scientific study of the role of the brain in aesthetic experience and artistic creation. Nevertheless, the context in which aesthetic experience should ultimately be understood is the cultural context that is rich with the enjoyed meanings that characterize our encounters with art. As Dewey's naturalism rightly posits, the biological life of an organism is continuous with this cultural environment. As such, neuroaesthetics has a significant contribution to make to the understanding of aesthetic experience writ large. Yet this contribution must be tempered with the knowledge that culture cannot be reduced to the functions of the brain, and that art and aesthetic experience are stripped of their most worthwhile contributions to human experience if the richness of the cultural context in which art is enjoyed is sought to be exhaustively explained by the biological substrate from which it emerges.

Notes

1 Zeki 2001, p. 52.
2 This, in fact, is how Dewey's aesthetic theory was received by his critics Stephen Pepper and Benedetto Croce. For more on the Pepper-Croce thesis see Croce 1948, pp. 203–7; and Pepper 1951, pp. 369–90.
3 Dewey 1981a, p. 10. Hereafter, *LW1*.
4 Randall 1944, p. 356.
5 *Ibid.*, p. 357.
6 *Ibid.*, p. 358.
7 Levine 1944, *Naturalism and the Human Spirit*, pp. 184–5.
8 Randall 1944, p. 369.
9 Dewey 1981b, p. 30. Hereafter, *LW12*.
10 Rosenthal 2004, pp. 158–9.
11 Dewey *LW12*, p. 72.
12 *Ibid*, p. 48.
13 *Ibid.*, p. 49.
14 *Ibid.*, p. 52.
15 *Ibid.*
16 *Ibid.*, p. 72.
17 Dewey *LW1*, p. 268.
18 *Ibid.*, p. 269.
19 Onians 2007, p. 147.
20 Dewey, 1987, p. 278. Hereafter, *LW10*.

21 Skov and Vartanian 2009, p. 2.
22 Brown and Dissanayake 2009, p. 45.
23 For two instances of pioneering work in neuroaesthetics that take an art-centric approach, Ramachandran and Hirstein 1999, pp. 15–51; and Zeki 1999.
24 Chatterjee 2010, p. 54.
25 Zeki 1998, p. 72.
26 *Ibid.*, p. 76.
27 *Ibid.*, p. 95.
28 Conway and Rehding 2013, p. 3.
29 Kawabata and Zeki 2004, p. 1699.
30 *Ibid.*, p. 1702.
31 Conway and Rehding 2013, p. 4.
32 Brown and Dissanayake 2009, p. 44.
33 Dewey *LW10*, pp. 16–17.
34 Nadal and Skov 2013, p. 5.
35 See Dickie 1964, pp. 56–65.

References

Brown, S., and Dissanayake, E. (2009), "The Arts are More than Aesthetics: Neuroaesthetics as Narrow Aesthetics," in *Neuroaesthetics*, Martin Skov and Oshin Vartanian (eds). Amityville, NY: Baywood Publishing Company.

Chatterjee, A. (2010), "Neuroaesthetics: A coming of age story," *Journal of Cognitive Neuroscience* 23: 1.

Conway, B. R., and Rehding, A. (2013), "Neuroaesthetics and the Trouble with Beauty," *PLOS Biology* 11: 3.

Croce, B. (1948), "On the Aesthetics of Dewey," *Journal of Aesthetics and Art Criticism* 6(3): 203–7.

Dewey, J. (1981a), *The Later Works of John Dewey, 1925–1953, Volume 1: Experience and Nature*, ed. Jo Ann Boydston. Carbondale, IL: Southern Illinois University Press.

—(1981b), *The Later Works of John Dewey, 1925–1953, Volume 12: Logic: The Theory of Inquiry*, ed. Jo Ann Boydston. Carbondale, IL: Southern Illinois University Press.

—(1987), *The Later Works of John Dewey, 1925–1953, Volume 10: Art as Experience*, ed. Jo Ann Boydston. Carbondale, IL: Southern Illinois University Press.

Dickie, G. (1964), "The Myth of the Aesthetic Attitude," *American Philosophical Quarterly* 1(1): 56–65.

Kawabata, H., and Zeki, S. (2004), "Neural Correlates of Beauty," *Journal of Neurophysiology* 91: 1699.

Levine, T. (1944), "Naturalism and Sociological Analysis of Knowledge," in *Naturalism and the Human Spirit* ed. Yervant Krikorian. New York: Columbia University Press.

Nadal, M., and Skov, M. (2013), "Introduction to the Special Issue: Toward an interdisciplinary neuroaesthetics," *Psychology of Aesthetics, Creativity and the Arts* 7(1): 5.

Onians, J. (2007), *Neuroarthistory: From Aristotle and Pliny to Baxandall and Zeki*. New Haven, CT: Yale University Press.

Pepper, S. (1951), "Some Questions on Dewey's Esthetics," in *The Philosophy of John Dewey*, ed. Paul Arthur Schilpp. New York: Tudor Publishing Company, pp. 369–90.

Ramachandran, V. S., and Hirstein, W. (1999), "The Science of Art: A neurological theory of aesthetic experience," *Journal of Consciousness Studies*, 6(6–7): 15–51.

Randall, Jr., J. H. (1944), "The Nature of Naturalism," in *Naturalism and the Human Spirit*, ed. Yervant Krikorian. New York: Columbia University Press, p. 356.

Rosenthal, S. B. (2004), "Pragmatic Naturalism, Knowing the World, and the Issue of Foundations," in *Dewey, Pragmatism, and Economic Methodology*, ed. Elias L. Khalil. New York: Routledge.

Skov, M., and Vartanian, O. (2009), "Introduction: What is Neuroaesthetics?" in *Neuroaesthetics*, Martin Skov and Oshin Vartanian (eds). Amityville, NY: Baywood Publishing Company.

Zeki, S. (1998), "Art and the Brain," *Daedalus* 127(2): 72.

—(1999), *Inner Vision: An Exploration of Art and the Brain*. New York: Oxford University Press.

—(2001), "Artistic Creativity and the Brain," *Science* 293: 5527.

Part Two

Reconstructing Neuroscience and Philosophy

5

Descendants of Pragmatism

Reconciliation and Reconstruction in Neopragmatism, Neurophilosophy, and Neuropragmatism

Tibor Solymosi

Pragmatism's historical relationship to the sciences (especially those of life and mind) has been neglected, if not undercut in its Rortian revival of the past three and a half decades. This neglect has many roots, from the linguistic turn in professional philosophy to the Kuhnian and postmodern subversion of scientific authority. A thorough exegetical analysis of this history is not the aim of this chapter.[1] Rather, my goal is to tell a story of pragmatism and its attitude toward the sciences. I do not intend this story to be fictional or duplicitous; nor, however, do I claim it to be *the* story of pragmatism. My hope is that this story suggests shortcomings to how classical pragmatism is contemporarily considered, how contemporary (neo)pragmatism is related to classical pragmatism, and how contemporary scientific philosophy, especially neurophilosophy, relates to pragmatism, old and new.

I continue a line of thought argued earlier[2] that neopragmatism is not the only resurgence of pragmatism in the past few decades: that neurophilosophy is an implicit reappearance of integral aspects of classical pragmatism. To illustrate the historical trajectory from a pragmatism enthusiastically friendly to science to a contemporary pragmatism concerned largely with social and moral life without much if any interest in contemporary science alongside a scientific philosophy with little interest in social and moral life, I focus on the conception of experience at work in neopragmatism and neurophilosophy.[3] The chasm that appears between neopragmatism and neurophilosophy resembles the dualism between value and fact that classical pragmatism sought to undermine. Central to the demolishing of dualisms for classical pragmatists like John Dewey is an evolutionary standpoint that conceives of experience as organism–environment transaction.[4] This transactional conception of experience is missing in both neopragmatism and neurophilosophy. A significant number of contemporary pragmatists who claim to endorse this transactionalist account of experience have yet to draw on the resources available in the sciences of life and mind—sciences to which the classical pragmatists were early contributors. The relatively few exceptions are the philosophically-trained pragmatists who have engaged with the neurosciences, or the neuroscientifically-trained researchers who have engaged with pragmatism: the set of thinkers I have called *neuropragmatists*.[5]

Neuropragmatism, it follows, is a way forward for both pragmatism and neurophilosophy (or scientific philosophy, more generally). The story told in this chapter is a story of how neuropragmatism can provide philosophical attention to issues involving the neurosciences. This story begins with a very brief introduction to Dewey's conception of experience as organism–environment transaction, emphasizing the role of scientific activity within experience, particularly as a socially transformative phase of experience. This view of science as socially transformative is shared by the logical positivists and empiricists with whom Dewey and other pragmatists such as Sydney Hook and Morris Cohen were engaged in the Unity of Science movement.[6] But what pragmatists and logical empiricists believed about scientific activity was not the same. For the underlying conception of experience was significantly different. The consequences of this difference prove to be significant for the very idea of a science of experience. After I review the differences between the pragmatists' radical empiricism and the logical empiricists' Humean or sensationalistic empiricism, I highlight how the pragmatist conception failed to take hold for both neopragmatism and neurophilosophy. This failure is at the root of the major conflict with which philosophy and science are faced today: how to *reconcile* our common-sense view of ourselves as conscious, free, and morally responsible selves with the growing scientific view that there is no place for selves, responsibility, freedom, or consciousness in reality (which is wholly material (so it is claimed)).

Neuropragmatism is well situated to evade this dilemma of reconciliation by rejecting the conception of experience that supports it. By modifying the classical pragmatist view of experience as organism–environmental transaction, I argue that neuropragmatism offers us the way forward for *reconstructing* common-sense experience via scientific activity in an ever ameliorative fashion. To conclude this chapter, I briefly consider the work of neuroscientist and popular author, David Eagleman. His recent book, *Incognito*, offers neuroscience as the apparently definitive science of experience. At times, he comes close to doing some reconstructive work, but as the rest of this chapter aims to develop, he falls into the reconciliast trap. I suggest a way out through reconstruction.

On experience: From Descartes to Dewey

When a person speaks about experience, he or she may have any number of things in mind. When philosophers speak about experience, they often have something very specific in mind—after all, if philosophers strive for anything, it is for a high degree of specificity. Unfortunately, as Marianne Janack has recently noted,[7] philosophers since at least early modernity have been using a conception of experience that is rather alien to the everyday uses with which most people are familiar. Indeed this confusion is at the heart of Dewey's initial attempt to save the word for philosophy and his later lament that he should abandon the word in favor of *culture*.[8] This sort of frustration is also at play with Richard Rorty's rejection of the word from his philosophical vocabulary. This is not to say that Dewey would have approved of Rorty's linguistic turn. Rather that Dewey's transactional view of experience was too often taken in a sensationalistic vein (I elaborate on this vein momentarily). Dewey's conception of culture—like experience—was—to put it in contemporary terms—embodied, extended, and enactive, as well as symbolic-linguistic. Rorty's view, however inspired by Dewey's frustration, seems to take up a specific linguisticism that has no place for the embodiment, extension, or enaction found in Dewey's transactionalism. To appreciate this trajectory through philosophy, we first need an appreciation of the orthodox philosophical conception of experience.

Beginning in antiquity, philosophers have distinguished between appearances and reality. Reality, on this view, is fixed and permanent; appearances can and do change, often without reason. Moreover, what humans experience are appearances of reality. When we talk about having knowledge about reality what we mean is that our appearances correspond with what is actually the case. When our experiences are faulty, they are just mere appearances that provide us with no confidence as to whether they represent reality as it truly is. Philosophers have proposed numerous solutions historically to this problem of knowing. With the rise of modern science, there was a sense that humans were starting to get a handle on reality given our increase in the ability to manage with greater stability the otherwise precarious world of nature.

This confidence was not without its philosophical problems. Descartes was not comfortable with the appearances of success in science. He wanted absolute certainty. To achieve it, he developed his method of radical skepticism: he would doubt all he could, beginning with how things appeared, from sensory data to what he had been taught—to even his own being. Only what he could not doubt would be considered absolute. Since he could not doubt that he was, in fact, doubting, he famously concluded, "I think, therefore I am." From this deceptively simple conclusion, Descartes inferred that he was not his body, and therefore no mind is identical with its body. From this mind–body dualism, there was an inevitable chasm between the immaterial mind and the material world. How was it possible for these two sorts of things—mind and body—to interact when they are initially conceived in diametric terms? Descartes was forced to postulate a benevolent deity whose gift of reason to humans enabled humans to deduce the fixed truths from which empirical science could prosper.

Descartes' solutions did not convince many. But his articulation of the problem of knowing about an external world set the framework for Western philosophy. Locke denied the Cartesian emphasis on reason, famously advocating his blank-slate theory of experience—but not without a deity to set things up for humans to inevitably know the external world. Hume took this empiricism to his logical end: to a skepticism not only about the external world but also about the very self Descartes believed indubitable.

This Humean empiricism held that there were experiences of sensations or sense data. What this data referred to, where it came from, was impossible to know for sure. For whatever this experiencer is, what was experienced are the bits of data and their relations to each other. The experiencer, in effect, experienced nothing but sensations or ideas.[9] There was no getting around these sensations to reality itself. Certain knowledge was impossible. Of course, this conclusion had little bearing on how humans actually lived in their day-to-day lives. And it certainly did not stop Hume from living his practical life.

It did, however, wake Kant from his dogmatic slumber. Kant's solution to the problem of knowledge—the details of which take us well beyond the scope of this chapter—aimed to bridge the gap between Cartesian rationalism and Humean empiricism. In doing so, it nevertheless recapitulated the ancient divide between things-in-themselves (which we can never experience and thereby never scientifically know) and that which is experienced (and thereby available to scientific investigation). Among the things-in-themselves, Kant included God, the soul, and free will. In short, there can be no empirical science of experience itself. That which makes experience possible—God and a soul on the conceptions of Descartes and Locke—are not available to experience nor to reason for investigation.

This barrier between appearance and reality, between mind and world, continues to bedevil many philosophers today. It also frames much scientific work, especially in the sciences that claim to address the experiential itself, despite Kant's conclusions. Neuroscientists especially are eager to talk about a science of consciousness, of mind, or even of the soul,[10] but often they do so only to conclude that such things are illusory. On the broadly Cartesian framework on which much philosophy and science operate, such conclusions seem inevitable.

Yet how did philosophers and scientists get to the point at which they were confident in scientifically addressing questions that were not long ago too ephemeral to reach by the methods of science? On the story I am telling here, Darwin's theory of evolution by means of natural selection offered humans a means of conceiving the uniquely human as continuous with—and not in opposition to—the natural and physical world. Among the first philosophers and scientists to recognize this import of evolutionary thinking were the classical pragmatists: Peirce, James, and Dewey.

Dewey argued in his seminal essay, "The Influence of Darwinism on Philosophy," that what Darwin's theory does is open the gates of the garden of life to the methods of science through which humans can begin to inquire experimentally into mind, meaning, and morals.[11] Dewey would later expound on this evolutionary continuity through his conception of experience as a nested hierarchy of physical, psychophysical, and mental plateaus or phases of nature.[12] This Darwinian view conceived of experience as not only central to philosophy and science but also in a radically different way from Descartes, Hume, and Kant.

The pragmatists rejected the dualisms between mind and body or mind and world that brought about the epistemological quagmire of modern philosophy. Mind grew out of body, hence no problem of mind–body interaction. Mind grew out of bodily interaction with the world, hence no problem of knowledge of the external world. Dewey took this Darwinian perspective to reconstruct the Humean or sensationalistic conception of experience into a transactional conception, in which experience is *organism–environment transaction*.

Such a conception, however, seems as alien to most people's use of the word *experience* as the sensationalistic conception. Before I elaborate on this transactional conception, consider Dewey's description of his use of experience, from "Creative Democracy—The Task Before Us." He writes,

> If one asks what is meant by experience … my reply is that it is that free interaction of individual human beings with surrounding conditions, especially the human surroundings, which develops and satisfies need and desire by increasing knowledge of things as they are. Knowledge of conditions as they are is the only solid ground for communication and sharing; all other communication means

> the subjection of some persons to the personal opinion of other persons. Need and desire—out of which grow purpose and direction of energy—go beyond what exists, and hence beyond knowledge, beyond science. They continually open the way into the unexplored and unattained future. Democracy as compared with other ways of life is the sole way of living which believes wholeheartedly in the process of experience as end and as means; as that which is capable of generating the science which is the sole dependable authority for the direction of further experience and which releases emotions, needs, and desires so as to call into being the things that have not existed in the past. For every way of life that fails in its democracy limits the contacts, the exchanges, the communications, the interactions by which experience is steadied while it is also enlarged and enriched. The task of this release and enrichment is one that has to be carried on day by day. Since it is one that can have no end till experience itself comes to an end, the task of democracy is forever that of creation of a freer and more humane experience in which all share and to which all contribute.[13]

Clearly, such a conception of experience is quite different from the sensationalistic view. But it is not as alien to the transactional view of organism–environment development. For that very simple transaction is at the heart of Dewey's description here: humans *interacting* with their human environment, which includes but is not limited to other humans. Furthermore, this view resonates more strongly with the quotidian sense that non-philosophers have. We share experiences; experiences are meaningful; experience can be enriched; experience is of other persons and of the world. The sensationalistic conception has difficulty connecting ideas, sense data, and logical operations to such uses of experience—to what Dewey referred to as *lived experience.*

To account for lived experience scientifically is proving to be an enormously difficult problem for philosophers and scientists—especially those cursed by creeping Cartesianism.[14] Pragmatists have a role to play in cleaning up the conceptual mess and further advancing inquiry. However, contemporary pragmatists have fallen into their own quagmire over the word *experience*. As my opening remarks indicate, among those who endorse the transactional view of experience most fail to utilize contemporary neuroscience—part of the "knowledge of conditions" about which Dewey wrote—while those pragmatists who reject transactionalism because they reject experience as philosophically cogent are depriving themselves of the means for doing the ameliorative work to which they believe philosophy ought to contribute. Such pragmatists become engaged in the project of reconciliation because they, like neurophilosophers, never fully escape the sensationalistic conception of experience.

To more fully appreciate the radical turn that neuropragmatism presents to neopragmatism, neurophilosophy, neuroscience, and classical pragmatism, I first draw on the contrasting conceptions of experience and connect them to a Deweyan/(neuro)pragmatist conception of experimentalism. In which scientific activity is the purposeful, deliberate, intentional touching, disturbing, disrupting, and sullying of a situation to effect novelties. The novelties that emerge from the creative activity of experimentation are tools and recipes—guides for action—that are as value-laden as they are constitutive of value. This experimentalism is key to Dewey's project of reconstruction; it, however, is just what has been lost in the resurgence of pragmatism in the last 30 years.

If we take seriously the conception of experience as organism–environment transaction or, more simply, *life-function*,[15] then we must grant the following:

1. There was a time in the past prior to:
   a Human experience
   b Animal experience
   c Any experience.
2. There are forms of experience entirely remote in time and space from human experience, yet, given evolutionary continuity with nature, there are some general traits to experience.
3. Among such traits are three integral ones:
   a All experiences are at least *had* or *undergone*—but not necessarily *known*.
   b Experience as life-function follows a rhythm, based in the thermodynamics of equilibrium regulation, out of which grow circadian rhythms and patterns of anticipation/consummation as we see in the cases of food and sex.
   c From this rhythm, the general pattern of *adjustment* emerges, viz., an organism *adapts* itself to its environment while *altering* its environment to itself.[16]
4. Continuity is not identity, therefore each form of life may share some aspects or forms of experience with other forms yet nevertheless maintain its own idiosyncrasies.
5. Human experience is not only biological but cultural, our environments are filled with symbols that afford us greater opportunities for action—for experiences not only had but those which are *known*.

For humans, knowledge is the honorific title we give to beliefs—habits of action—that have thus far proven to be true, reliable guides for getting about our bio-cultural environments. The methods by which beliefs get fixed, as Peirce taught us, do not always guarantee knowledge; but there is one method that has developed to be self-correcting, in part because of the hard-won recognition that our beliefs are often mistaken and in need of revision. This general method is what Peirce called the scientific and what James and Dewey also called the experimental. Since this method is a phase of experience as life-function, the above traits apply to it. For present purposes, it must be emphasized that there is no one single experimental method. Rather, there are methods that are experimental because they have been designed by human inquirers who strive toward the common end of deliberately resolving a problematic situation through controlled adjustments within that situation.[17] For good reason, then, Dewey saw one of the central jobs of philosophy as the inquiry into inquiry: the study of the methods of diverse subject-matters with an aim of developing a *general* theory of inquiry that would assist in the development of new experiments, not least because of the hope that various fields may have something useful to offer one another.

In order to illustrate that neuropragmatism is a way forward for pragmatism, especially if it is to engage the sciences of experience, so to speak, my tale now turns to the descendants of pragmatism, their underlying sensationalistic conception of experience, the consequences for understanding science, and how this modern dualism remains a problem for philosophers and neuroscientists today. After this discussion, I return to the Deweyan conception of experience as organism–environment transactionalism and upgrade it given contemporary scientific insights.

Experience in neopragmatism and neurophilosophy

Colin Koopman notes that for Rorty and his followers the concept of experience fails to do the anti-foundationalist work James and Dewey thought it could.[18] In fact, the desire to achieve this anti-foundationalism results in the elimination of the concept of experience because experience is seen as the ultimate arbiter—the foundation—of all further claims, philosophic and scientific. The elimination of experience also occurs in neurophilosophy, perhaps most notably in the works of Paul and Patricia Churchland. The line of thought is that the Cartesian theory of mind has no empirical, viz. scientific or factual, support. Therefore, our experience of *being immaterial souls* is illusory—a claim which too often is simply truncated to the claim that *experience* (or some variety of it, e.g. consciousness or mind) is illusory.

In both cases, these descendants of pragmatism share a desire to reject Cartesianism. For both, the rejection consists in the epistemological dismissal of introspection and claims to certainty—this is central to the pragmatist ancestry. Similarly, there is the ontological rejection of substance dualism. For a neopragmatist like Rorty, this rejection goes so far as to reject metaphysics entirely. He is not interested in talking about concepts like Being, Reality, or Truth; he no longer finds them useful. Neurophilosophers, however, do hold onto the metaphysical import of science. Contrary to Rorty's antirealism, there is the underlying scientific realism of neurophilosophers like the Churchlands.

Consider Rorty's Sellarsian imperative that "all awareness is a linguistic affair."[19] There is no awareness (or experience) prior to the use of language. The neurophilosopher does not reject the priority of language here, but she does want to know how this linguistic awareness is possible in the first place. Her story, briefly and crudely put, is that within a human's brain what is *really* going on is a complex network of synaptic firings of such and such a pattern. In other words, the linguistic affair is really something physical, not something as ephemeral as souls conversing with one another.[20]

This disjunction between these two descendants of pragmatism leads to what Sellars referred to as the clash of the images of science and common sense (what Sellars called the manifest image). I return later to this clash and the attempts to deal with it in neopragmatism and neurophilosophy. For the present, I turn to an underlying commonality between the two, namely the inability to escape experience.

The ancestry of neopragmatism and neurophilosophy consists of the pragmatism of James and Dewey *and* the pragmatism of logical positivism and empiricism found in Carnap, Quine, and Sellars.[21] While these two pragmatisms rallied against modern empiricism, only one was radical enough to evade the problem of a mind/world barrier. This conception of experience—James and Dewey's—was inherited by neither neopragmatism nor neurophilosophy. The sensationalistic conception of experience remained in Quine and found its way to these two descendants of pragmatism.

This sensationalism is well-illustrated by the now long-popular metaphor that the mind/brain is like a computer, in which the mind is the software that runs on the hardware of the brain. For the moment, never mind that this view is antiquated.[22] I want to focus on how this mechanical metaphor works.[23] For the computer, there are inputs that come from the keyboard and mouse, and outputs that come from the screen, speaker, or printer. For Quine, there are the sensory inputs that come at the boundary of the skin, and then the consequent output in behavior, such as a speech or bodily act. Notice how well this computer sensationalism fits the neopragmatist and the neurophilosophical views. The former sees language like a code for computer software: one writes, speaks, or types an input, and gets output in return (which becomes new input to be processed). The neurophilosopher is less interested in the software and seeks to understand how the hardware does the job. What we have then is a threefold divide between mind (software), body (hardware), and the world. The modern conception of experience pragmatists are so adamant to escape is simply reiterated in neopragmatism and neurophilosophy.

The projects of reconciliation and reconstruction

The major figures among the classical pragmatists—Charles Sanders Peirce, William James, John Dewey, and George Herbert Mead—were, to varying degrees, immersed in the sciences of their day. From Peirce's experimental work in geodesy to James's *Psychology*, and Dewey's educational work and proximity to major advancements in physics, biology, and developmental psychology[24] to Mead's sociology, the classical pragmatists not only found interesting and valuable the insights of science, they also took up its methods as exemplifying a general orientation, and a pattern—a theory—of human inquiry. Especially in the instrumentalism of Dewey, the consequences of science were of significant import for everyday lived experience. These consequences were understood in at least two respects. First, the social, material, and industrial transformations that scientific advancements brought about were and continue to be undeniable for human experience: much has changed for humans in the past two centuries. From the industrial revolution to the informational, from the germ theory of disease to its genetic underpinnings, the tools developed for and made possible by scientific activity have had and will continue to have significant consequences for human life. Secondly, there is the consideration of the facts and values that make scientific activity possible as well as how that activity itself along with its results transform beliefs about both fact and value.[25]

For Dewey, the activity of philosophical reflection focused on what to make of scientific activity in light of how we conceive of ourselves[26] was at the heart of the project of *reconstruction*. Dewey contrasted this view of philosophical activity with the project of *reconciliation*. The central difference between these two philosophical projects is the two different conceptions of experience and the subsequently different conceptions of science on which these projects are based.

Consider two of Dewey's passages on these projects. First, from *The Quest for Certainty*, Dewey, writes:

> The problem of reconciliation arises and persists for one reason only. As long as the notions persist that knowledge is a disclosure of reality, of reality prior to and independent of knowing, and that knowing is independent of a purpose to control the quality of experienced objects, the failure of natural science to disclose significant values in its objects will come as a shock.[27]

And here is Dewey on reconstruction, in *Reconstruction in Philosophy*:

> In short the problem of reconstruction in philosophy, from whatever angle it is approached, turns out to have its inception in the endeavor to discover how the new movements in science and in the industrial and political human conditions which have issued from it, that are as yet only inchoate and confused, shall be carried to completion. For a fulfillment which is consonant with their own, their proper direction and momentum of movement can be achieved only in terms of ends and standards so distinctively human as to constitute a new moral order.[28]

The problem of reconciliation comes to this: so long as we believe that science discloses reality as it is independently of us and undisturbed by human activity, we will forever be bewildered and troubled by the experience of value and quality in our lives. The problem of reconstruction, however, comes to this: once we recognize scientific activity as occurring in a specifically human context that is qualitative and value-laden, dependent upon and responsive to qualities and values as much as it is reliant upon and productive of facts, we take on a responsibility for deliberate action toward our ideals that are afforded by the facts of the case. This in turn requires both reflection upon our values to determine whether they are valuable, as well as reflection upon which inquiries to pursue next as they are more or less likely to effect amelioration. However, the project of reconciliation in the face of the purportedly absolute facts of an efficiently causal and strictly quantitative and non-human world is simply dumbfounded when it comes to ideals, values, and qualities. These differences in conceptions of experience, science, and the projects of reconciliation and reconstruction are key to understanding the underlying relationship between neopragmatism and neurophilosophy. Despite seemingly deflationary views of experience as philosophically cogent, these two descendants of pragmatism remain entangled with the problems of sensationalistic empiricism.

Reconciliation in neopragmatism and neurophilosophy

More and more neurophilosophers are finding inspiration from or affinities with pragmatism, especially Dewey's work.[29] Whether these neurophilosophers understand Dewey properly is not my main concern. What is my concern is noting that one of the reasons pragmatism lost its grip in professional philosophy is that many of the students of Dewey and Mead did not continue in professional philosophy. Rather they went into experimental science, especially the social sciences. Their influence, as well as James's, also carried through in the sciences of life and mind.[30] As Jay Schulkin has noted, psychobiology—a precursor to contemporary evolutionary cognitive and behavioral neurobiology—has among its founders James and Dewey.[31] The consequences of pragmatism for experimental science were substantial, though their pragmatist origins were largely unrecognized, for scientists tend to be unconcerned with the history and philosophy of their discipline. When analytic philosophers began (re)turning to the sciences, especially in the form of neurophilosophy, they (re-)encountered pragmatic themes and ideas. However, they encountered them with an impoverished conception of experience as a strictly linguistic, if not representational affair—an impoverishment that subsequently bled into their conception of science.

To be clear, neopragmatism and neurophilosophy are the children of the congress of classical pragmatism and mid-century analytic philosophy. As is the case with any set of siblings, they share traits common to both parents as well as inherit traits unique to one parent and unexpressed in a sibling. While of greater worth than the insular, a priorism of analytic philosophy, the cash value of either resurgence of pragmatism is not as great as its parent classical pragmatism. Neopragmatism focused on ideals and values and how best to speak of them, but it paid no heed to experimental methods of achieving them. Conversation has its place as does literature, but neither are—nor should they be—the be-all and end-all of philosophy. The demotion of the authority of scientific activity that is associated with the postmodernism of neopragmatism did not carry through to neurophilosophy. Its obsession with neuroscientific fact, however, did push away any serious consideration of values or ideals. In fact, there was no place for such talk. Where neopragmatism sought to eliminate scientific authority, reducing it to literature like all the rest, neurophilosophy sought to eliminate anything but science as being capable of telling us about the world.

To reiterate, common to both impoverished pragmatisms is their conception of experience. Their conception, to be more specific, is not just some species of the sensationalistic sort of Hume but a variety of Sellars, for whom "all awareness is a linguistic affair." Rorty's Sellarsianism becomes a matter of reconciling what we know or think we know with what we hope. Instead of making use of pragmatic experimentalism, Rorty's use of Kuhn to reject scientific authority (recall the authority gained from a third-person public perspective) leaves him with what we might call *eliminative idealism*.[32] Neurophilosophers embraced scientific authority and left nothing to hope for beyond a final theory of the nervous system that would do away with our self-conception, firmly establishing it as illusory. This is eliminative materialism—coined by none other than Rorty himself.[33] One theory replaces another; one language game eliminates or makes superfluous the other.

The most recent construal of the project of reconciliation comes from the mid-twentieth century philosopher, Wilfrid Sellars. In an influential essay, "Philosophy and the Scientific Image of Man," Sellars argued that the job of philosophy was to see how things hang together in the broadest possible sense.[34] Most strikingly, philosophy's greatest job is to show how two competing conceptions of humanity can hold together. These two views Sellars called the scientific image and the manifest image. The latter corresponds to our self-conception as conscious, free, and morally responsible selves. Sellars's conception of science as value-free and objective, of metaphysically unveiling the true structure and nature of reality, is a common and intuitive[35] understanding of science, particularly among philosophers and scientists alike. The status of the manifest image is problematic for the scientific realist.[36] On the one hand, the image is manifest in that it is obvious or clearly understood to any mind. On the other hand, this image is a manifestation, a mere appearance, an apparition—not *really* real. Such a view that our obviously value-laden and valuable experiences as conscious, free, and morally responsible selves are really nothing but mirages is a view that threatens human dignity.[37] Indeed, our self-conception is what has traditionally been advocated by the greatest humanists of history. Their defense of the human spirit and human dignity is illustrated through literature in all its forms.[38] The problem of reconciliation is that the two images set themselves in competition with each other as, to use Sellars's words, "the true, complete, and final" view of humanity. Insofar as these images are diametrically opposed, how they hang together is unclear and has been a subject of debate for philosophers.

Two possibilities, however, have emerged on how to reconcile the two images. Both possibilities share the similar pattern of reducing or eliminating one image to the other. Many neurophilosophers have advocated the position of eliminative materialism, which holds that our folk psychological concepts (i.e.

the manifest image, viz. our commonsensical view of ourselves as conscious, free, and morally responsible selves) will be replaced by and ultimately deemed as irrelevant in light of a full and mature neuroscience. Such a complete science leaves nothing left to be explained—nothing more is to be said about consciousness, freedom, etc. If neurophilosophers eliminate the manifest image in favor of the scientific image, then the only other alternative is to eliminate the scientific image in favor of the manifest image. This is the move neopragmatists like Rorty have made. By rejecting the authority of science to disclose the world or reality, Rorty argues that science is just another means of producing literature that has as much or as little claim to authority as any other form of literature. Since Rorty and his followers are mostly concerned with humanistic ideals like democracy and freedom, it is appropriate, at least for my purposes here, to refer to them as eliminative idealists. After all, for Rorty and Sellars, all experience is nothing other than a linguistic affair. Contrast this view to the alternative, that all experience is a material affair, and we see the import of referring to the scientific eliminativists as materialists and the postmodern-linguistic eliminativists as idealists.

In either case, science is seen as claiming to reveal or disclose reality independently of scientific activity. The eliminative materialist believes this to be true of science; the eliminative idealist believes this is what science believes itself to do but rejects any validity to the claim. If we are to follow Dewey's advice that the way out of the problem is to recognize that all knowledge is the product of the activity of knowing, that all scientific claims are the product of scientific activity, then what happens to these popular positions on how to regard science?

The answer lies in what Dewey called the project of reconstruction. He saw that many people saw a tension or conflict between science and common sense. However, he did not fall into the traditional philosophical mistake of believing that there is a world knowable to humans without human activity. Dewey described the project of reconstruction by framing scientific activity, its developments, and its products in their social, political, and cultural context. The modifications made and being made possible by science are not to be taken as givens or taken lightly.[39] Philosophical reconstruction aims at guiding scientific activity in light of our cherished ideals as well as adjusting those ideals through criticism effected by scientific data. This embedding of scientific activity and its products in culture resists the pull of eliminative materialism and eliminative idealism.

Eliminativism of either sort is the result of the neglect of the experimental method. This neglect divides experience into facts and into values, or mind and body, or the mental and the physical. Pragmatic reconstruction rejects such a division and seeks to utilize the continuities of mind and body and the entanglements of fact and value toward ameliorative action. This work has yet to be completed, especially with regard to recent scientific advances.[40] Many neurophilosophers today are recognizing the limitations of the reconciliatory view; others have reconciled themselves to it. Among the latter is the one who coined the term neurophilosophy, Patricia Churchland. In her recent book on morality, she writes that neuroscience, along with other sciences like anthropology, can help us learn "to cope with social problems," given our predicament as products of evolution in an often unfriendly world.[41] While Dewey would certainly agree that the sciences can help us deal with a world that is both precarious and stable, he would have gone for more than merely coping; he would have advocated a consummate meliorism. Such language, however, is not germane to Churchland. As many will recognize, talk of coping is the talk of Rorty.

Neuropragmatism and reconstruction: Moving forward

I believe we pragmatists can do better than this, especially when it comes to the consequences of neuroscience. In putting forth neuropragmatism, I aim to bring to bear the tools and lessons of Dewey and other pragmatists to a present situation in which neuroenthusiasts[42] are already talking about and working toward bringing about a neuro-society or neuro-culture that, they claim, promises to radically change the human condition. Neuroenthusiasts like Zack Lynch[43] and Lone Frank[44] speak of neuroscience as the next big revolution of human civilization, akin to the Copernican or the Industrial revolution. Frank speaks of neuroscience as providing the so-called mind/brain the ability to know itself absolutely.[45] Others still believe that we are simply on the cusp of a radical revolution based in the brain. To review the growing literature of near fanaticism about the brain and its promise would take us well beyond the scope of this chapter. However, to combat the reconciliation of neopragmatism and neurophilosophy, an alternative conception of experience and science is required. The pragmatist conception of experience as organism–environment transaction not only evades the problems of sensationalistic empiricism, it also provides a conception of scientific inquiry that affords reconstructive amelioration of our lived experience through science.

The step to take here is to provide an overview of key themes to Dewey's philosophy of science that directly combat the reductionism of contemporary neuroscience (especially its enthusiasts). I contend that Dewey's (unappreciated) distinction between the empirical and the experimental strikes at the heart of the conflict of the projects of reconstruction and reconciliation.[46] Since my claim rides on the as-yet-unelaborated pattern of inquiry, I begin with Dewey's characterization of it. The general pattern of inquiry that pervades much of Dewey's thought is clearly stated by Dewey in five steps:[47]

i a felt difficulty;
ii its location and definition;
iii suggestion of a possible solution;
iv development of reasoning of the bearings of the suggestion;
v further observation and experiment leading to its acceptance or rejection; that is, the conclusion of belief or disbelief.

One method of inquiry that Dewey discusses is the empirical method where "inferences depend upon habits that have been built up under the influence of a number of particular experiences not themselves arranged for logical purposes."[48] Since there is a lack of such arrangement, this method is haphazard; it works but without guarantee of success or productivity. Dewey contrasts this method to the experimental or scientific method of inquiry.

"Scientific method," Dewey writes, "replaces the repeated conjunction or coincidence of separate facts"—i.e. the empirical method—"by discovery of a comprehensive fact, effecting this replacement by *breaking up the coarse or gross facts of observation into a number of minuter processes not directly accessible to perception*."[49] This is the analytic movement of discriminating processes or patterns that prior to the inquiry are acting in unison without being recognized as such. Many philosophers and scientists today are very enthusiastic about reductive analysis. Dewey saw experimental method as "a conjoint process of *analysis and synthesis*."[50] Once inquirers have discriminated the processes or patterns at work and how they work together to perform the function(s) under examination, the scientific inquiry is only complete upon the assimilation of these new findings. What are assimilated are not just the new patterns produced by inquiry but the habits, beliefs, and patterns that were not under direct investigation but nevertheless provided a scaffolding for inquiry.

The difference between empirical and experimental methods is significant. More often than not many scientists and philosophers fail to distinguish between these methods.[51] What is worse is that they also neglect the synthetic movement of science. Since these misconstruals are at the heart of the shortcomings of neopragmatism, neurophilosophy, and neuroenthusiasm, I quote

Dewey at length here on the distinction between empirical method and the scientific:

> The change of attitude from conservative reliance upon the past, upon routine and custom, to faith in progress through the intelligent regulation of existing conditions, is, of course, the reflex of the scientific method of experimentation. The empirical method inevitably magnifies the influences of the past; the experimental method throws into relief the possibilities of the future. The empirical method says, "*Wait* till there is a sufficient number of cases"; the experimental method says, "*Produce* the cases." The former depends upon nature's accidentally happening to present us with conjunctions of circumstances; the latter deliberately and intentionally endeavors to bring about the conjunction. By this method the notion of progress secures scientific warrant.[52]

The pattern here is one of activity that strives to go on without difficulty, without breakdown, without disruption. Before there were inquirers, this activity proceeded in ebbs and flows without any conscious awareness or deliberate action. Any disruptions to the equilibrium of the system—a change in pattern—yielded mutation and evolution. Organism–environmental transaction adjusts. Experience develops first into an empirical method of organisms taking notice of regularities in nature, and then, eventually, into taking deliberate steps to manipulate or manage these regularities. This deliberate activity demarcates the experimental method. The resolution of disruption is the aim of inquiry. In the case of scientific inquiry, the disruptions are deliberate and for the sake of resolving difficulties through enlarging the reach or scope of experience.

To return to the competing conceptions of experience and science, the eliminativists are faced with finding some way of reconciling two opposing sets of beliefs. The first is the set of beliefs that were not under direct investigation and, indeed, made inquiry possible: it is the cultural scaffolding of inquiry, itself a product of previous cultural inquiries and biological adjustments—of previous experiences. The second set of beliefs is the outcome of the analysis of the immediate subject-matter under experimental scrutiny. The reconciliast treats each set as competing views that are fixed and final. The reconstructionist rejects the opposition and acts on the belief that these two views are dynamic, fluid, and continuous with one another (even if the continuity is not readily understood). Because of this shift in perspective, the reconstructionist does not believe experimentation to be over until the results of the specific inquiry are synthesized with the rest of the beliefs that afforded that inquiry in the first place. To illustrate the differences between the reconciliast and the reconstructionist,

I suggest using *reductionism* to refer to the reconciliast project, and *analysis/synthesis* for the reconstructionist. The difference is that the former operates on the assumption of opposition, while the latter seeks continuity.

A brief example of reconstruction: What to do with the neuroenthusiasm of Eagleman

Neuroenthusiasts are not the only ones whose interpretation of neuroscience needs philosophical attention. Neuroscientists themselves, especially authors of popular science books, deserve significant attention from neuropragmatists. Consider David Eagleman, the best-selling author of the best-selling *Incognito: The Secret Lives of the Brain*.[53, 54] In this book, Eagleman provides an informative introduction to much of the exciting insights that we have learned about how human brains and human mentation work.[55] Like Damasio, Ramachandran, and Sacks before him, Eagleman is eloquent and entertaining. However, the philosophical moves he makes, consciously or not, are intriguing, if not problematic.

Given my present focus on conceptions of experience and the limits of this chapter, I focus primarily on a couple of passages from the second chapter of Eagleman's book, *Incognito*, before gesturing toward how to deal with far more complex and controversial claims Eagleman makes later in his book regarding freedom, responsibility, and law. The second chapter's title itself is quite telling: "The Testimony of the Senses: What Is Experience *Really* Like?" The opening section—"Deconstructing Experience"[56]—begins with Eagleman's account of Ernst Mach's phenomenology of what we now call Mach bands. Imagine a series of uniformly colored strips that are placed next to one another in order of decreasing darkness. On the left you have a very dark strip and all the way on the right there's a very light strip. The strips between them are shades of grey, lighter than the one on the left but darker than the one on the right. When the strips "were placed side by side each appeared to have a gradient of shading," as Eagleman describes the illusion, "slightly lighter on the left side, and slightly darker on the right."[57] What Eagleman goes on to say after introducing the illusion is noteworthy. He writes:

> Now that you are aware of this illusion of "Mach bands," you'll notice it elsewhere—for example, at the corner where two walls meet, the lighting differences often make it appear that the paint is lighter or darker right next to the corner. Presumably, even though the perceptual fact was in front of you this

> entire time, you have missed it until now. In the same way, Renaissance painters noticed at some point that distant mountains appeared to be tinted a bit blue—and once this was called out, they began to paint them that way. But the entire history of art up to that point had missed it entirely, even though the data was unhidden in front of them. Why do we fail to perceive these obvious things? Are we really such poor observers of our own experiences?
>
> Yes. We are astoundingly poor observers. And our introspection is useless on these issues: we believe we're seeing the world just fine until it's called to our attention that we're not. We will go through a process of learning to observe our experience, just as Mach carefully observed the shading of the strips. What is our conscious experience *really* like, and what is it not like?[58]

On the one hand, in this remarkable passage, Eagleman rightly points out that once we understand the way the illusion/perception works we notice it everywhere. As the example of the Renaissance artists illustrates, we learn through our experience (i.e. our transaction with the environment) and adjust our methods accordingly in order to bring about new experiences (i.e. new ways of interacting with the environment). As satisfying as this may be, Eagleman seems to ignore the more pressing and more interesting question: *how are we such bad observers/experiencers of the mountains?* This is the sort of reconstructive movement that Dewey had in mind when he argued in the 1938 *Logic* that the conclusions of science should feed back into common sense and enrich lived experience.

Yet Eagleman, perhaps in his desire to be provocative, changes the game. Instead of continuing on about how reflective inquiry can modify and improve our experience,[59] he asks questions that bring back Cartesian egos and intuitions—just the stuff that Dewey sought to eliminate. Instead of embracing the experimentalism that is barely alluded to earlier in the passage, Eagleman falls back on a naïve scientific realism that eliminates the import of primary lived experience.

The standard neurophilosophical account would largely follow Eagleman here. Basically, we humans don't *really* know what the world is like because we don't even know what our experience is *really* like; we don't know what our experience is really like because I (or any first-person perspective) have only one method, introspection, to investigate my own experiences. The neopragmatist response to this is harder to catch given the protean nature of postmodern relativism, but I think it would go something like this: your science is all very fine and well, indeed it tells a good story, but insofar as my experience goes, I am only and you are only aware of what we can put into words, and since any talk of "really" or "reality" or any qualitative experience beyond that which I can put into words is inevitably bound to confuse rather than liberate, all I "really" think your science can tell me is how I should have my kitchen painted to bring about a tranquil effect for me—by which I mean I could say to myself or another, "Ah, isn't this nice and tranquil?"—as I have my coffee in the morning.

The neurophilosopher reconciles herself to a life of fighting off daily illusions. The neopragmatist reconciles himself to a life of repainting the house. The neuropragmatist can offer better.

The first step in a neuropragmatic reconstruction of Eagleman's passage here is to reject questions or concerns about what my experience is *really* like. When it comes to illusions like Mach bands, we simply accept that we do experience them that way. Once we give up on questions about how the world is independent of our activity, we can begin asking questions about how our activity works. From there, we can ask further questions, like "how can we improve how we work?"

Central to this approach is recognizing that failures are just what calls to our attention our misperceptions of the world. When Eagleman claims we are poor observers because we may misperceive or make some other mistake about the world, he does not appear to be curious about how or why such mistakes come to our attention. In other words, what is lacking here is an account of inquiry and thus a thorough experimentalism. We are not poor observers of the world; we are actors in a world that is changing partially in response to our actions. Our actions are indicative of our interests and values, and they are necessarily limited in scope and detail. Yet through our evolution as a species and as cultures, we have developed new ways of engaging the world—ways that often enrich our experience. Eagleman's example of Renaissance art illustrates such achievements, not failures as he seems to believe.

With regard to Eagleman's specific question: "what is experience *really* like?" the differences in approach come to this: Eagleman is asking, "what is experience like independently of experience? What is it really like to be me without me?"[60] The neuropragmatist asks "how does our experience work? How can we control the conditions—the variables and the parameters—of our engagement with our environment so that we can act in not only more productive ways but also in ways that enrich lived experience?"

Recall Dewey's description of experience in "Creative Democracy—The Task Before Us." He situates experience in terms of values, needs, desires, and knowledge of conditions. It is through evaluating the values, needs, and desires in light of the facts of the case that we determine further action as to whether we fulfill those ends. Through the growth of scientific activity and the knowledge it produces, we are afforded opportunities for improving our interactions with each other, our cultures, our worlds. Eagleman calls his position a reductive materialism. There is nothing reductive about the data he draws upon. Such data is liberating; it provides opportunities for humans, as Dewey put it, "to go beyond what exists … into the unexplored and unattained future."

We should be cautious about the use of neuroscience—just as we should be with any scientific data we use[61]—but we should not keep the scientific attitude—the fallibilism, the transactionalism, the process of inquiry—at arm's length precisely because new data opens up new vistas for new inquiries and new experiences.

Such a line of inquiry, however, is not limited to neuroscience. Nor is it entirely reliant upon the nervous system or the science of it. While there can be no reasonable doubt that the nervous system plays an integral and coordinative role in experience when understood as organic-environmental transaction, we must not neglect or downplay the rest of the organism and its environment. This means that neuropragmatists need to be able to work with others in anthropology, ecology, and physiology, for instance, to answer these questions.

The consequences of neuroscience are still unclear. They will undoubtedly be significant. As neuroscience advances, we pragmatists are in a unique position to navigate new waters with new tools because we have at our disposal a living, breathing conception of both experience and science that recognizes its historical context and its capacity for further growth.

Notes

1 However, the resources for such a claim must include Reisch 2005. Other important works for such an enterprise should include Janack 2012 and Misak 2012.

2 Solymosi 2011a, 2011b.

3 That is, until very recently with the rise of bio- and neuroethics (see Racine 2010 and Moreno 2012).

4 See Solymosi 2013a, 2013b, and Solymosi and Shook 2013b. These articles elaborate the situational unity of experience by drawing on recent work in evolutionary biology, where the symbol Œ is the primary evolutionary unit. We borrow that symbol to indicate the entanglement of organism-environment as an ongoing process of adaptation—or learning from experience. For my purposes here, the use of this symbol is not necessary.

5 See Solymosi 2011a, 2011b.

6 Reisch 2005.
7 See Janack 2012. Cf. Koopman 2007.
8 Dewey 1925 *Later Works, vol. 1*, p. 361. Hereafter *LW1*.
9 This view was called by Dewey "the spectator theory of knowledge," (1929 *Later Works, vol. 4*, p. 19. Hereafter *LW4*.) and has been referred to as the Cartesian Theater by Daniel Dennett (1991, pp. 107, 134–5).
10 Cf. Humphrey 2011.
11 Dewey 1910 *Middle Works, vol. 4*, p. 7.
12 Dewey *LW1*, p. 208, and 1938 *Later Works, vol. 12*, pp. 26, 30–1. See also Solymosi 2011b, pp. 352–7.
13 Dewey 1939 *Later Works, vol. 14*, pp. 229–30. Hereafter *LW14*.
14 See Solymosi 2013b, and Solymosi and Shook 2013b.
15 Here I draw from Dewey. On organism–environment interaction or transaction, see *LW1*, p. 12, and *LW14*, p.16. On experience as life-function see the final chapter of Dewey 2012.
16 Adjustment occurs when either adaptation or alteration does; adaptation does not imply alteration, nor vice versa. However, it is often, if not usually, the case that both forms of adjustment go on simultaneously. See Hickman 2001, p. 21; and 2007, pp. 144, and 200–1. See also Solymosi 2011b, pp. 259ff.
17 See Tschaepe 2009 on pragmatic considerations in scientific explanation, and 2011 on Dewey's conception of scientific explanation which evades the realism/antirealism debate.
18 Koopman 2007.
19 Sellars 1956; Rorty 1982.
20 Cf. Churchland 2013.
21 Churchland 1986; Reisch 2005; Misak 2013.
22 Indeed, Dewey saw the shortcomings early on in the 1930s and 40s, see Dewey 2012, p. 217.
23 Here I paraphrase greatly Janack's careful and thorough exegesis of Quine, Rorty, and Kuhn. See her 2012, especially pp. 26–54, and ch. 3, "Children and Other Living Computers," pp. 67–88.
24 See Dalton 2002.
25 See Putnam 2002.
26 Variously called common sense, primary experience, our self-conception, the humanist or manifest image, or folk theories, to adapt more recent labels. As I have been describing it, our self-conception is that we are *conscious, free, and morally responsible selves*.
27 Dewey *LW4*, p. 35.
28 Dewey 1948 *Middle Works, vol. 12*, p. 275. Hereafter *MW12*.
29 See Solymosi 2011a, b, and Solymosi and Shook 2013a, for review of this growing affinity.

30 Dalton 2002, pp. 278ff.

31 See Schulkin 2009.

32 See Solymosi 2011a.

33 Rorty 1970. Some may wonder whether I am arguing that Rorty is both a materialist and an idealist. I am not making a claim either way. What I am implying is that materialism and idealism need not be diametrically opposed to one another. Such a dichotomy is the result of the analytic inheritance. Readers familiar with Dewey will recognize that the project of reconstruction can be put in terms of materialism and idealism: the material serves as the means by which to achieve our ends-in-view, i.e. that which effects the ideal. See Dewey *LW1*, p. 201. Furthermore, Patricia Churchland has recently recognized the limitations of eliminativism because of the consequences of the very word. In an interview by Julian Baggini (2012), we learn: "She accepts that part of the reason people have been dismissive of her and Paul's position is that its name is misleading." So why call this eliminative materialism? [Churchland replies:]

> I can tell you why we did that. It was a conscious decision. The expression wasn't new with us. It came from Richard Rorty. Who is Richard Rorty? He was a big gun at Princeton. Who were we? Very little fish in this nothing place in Canada. And we had this very self-conscious conversation, where we said, well, look, what are we going to do with this? We talked about calling it revisionary materialism, and Paul said, look, if we introduce a whole new term here (a) people aren't going to recognize it, so they aren't going to read it, and (b) they're going to say who the fuck are these upstarts, and we will simply be dismissed. So we thought better to take something that's recognizable and go with it. In the end I think that was a mistake. I'd call it revisionary materialism if I had to do it all over again, I'd call it really nice guy materialism if I had that opportunity, I'd give it a really nice name (70).

Further comment regarding how apropos such remarks are for this chapter and volume on neuropragmatism is not for me to make. I do, however, wish to emphasize that my treatment of eliminative materialism is pragmatic insofar as this is how philosophers, scientists, and others have come to use the term. Besides, my hope is that the reconstructive turn I propose is something that Patricia Churchland finds nice enough—an ideal materialism, if you will.

34 Sellars 1963.

35 This is to say nothing more than many if not most people simply are inclined to believe this without any consideration of experience of the facts (cf. Peirce's "The Fixation of Belief" and the a priori method).

36 The following play on the *manifest* is mine and not to be considered a position Sellars himself would have endorsed.

37 Cf. Dennett 2008.
38 Krikorian 1944; Snow 1959.
39 Dewey *MW12*, p. 275.
40 Much of the reconstructive work that has been done since Dewey is largely exegetical in nature, and/or more focused on the political and the cultural than the scientific, let alone the neuroscientific. Mark Johnson and Jay Schulkin have done the most toward this reconstructive end in light of cognitive and behavioral evolutionary neurobiology.
41 Churchland 2011, p. 4.
42 See Solymosi 2011a.
43 Lynch 2010.
44 Frank 2009.
45 Following Tom Wolfe, not Hegel. *Ibid.*, pp. 7 and 300ff.
46 For more on my pragmatist views of inquiry, see Solymosi 2012a, 2012b, 2013a, and Solymosi and Shook 2013b.
47 Taken from Dewey 1910 *Middle Works, vol. 6*, pp. 236–7.
48 *Ibid.*, p. 293.
49 *Ibid.*, p. 296, emphasis in original.
50 *Ibid.*, p. 298, emphasis in original.
51 Indeed what distinguishes neuropragmatism from neurophenomenology is the former's emphasis on experimentalism, which is lacking in the latter.
52 *Ibid.*, pp. 299–300. It strikes me that much neuroscience is the result of comparing brain-damaged people's behaviors with non-damaged people's, and then looking for corresponding differences across the brains. In many ways, neuroscience has been far more empirical than experimental, at least with regard to humans. Rats, however, are an entirely different matter.

This is not to disparage the growing body of literature on clever experimental set-ups that do not rely on brain-damaged subjects. There is further cause for greater excitement and hope with the rise of *neuroanthropology*, which seeks to study "brains in the wild" (Lende and Downey 2012).
53 In noting that both Eagleman and his recent book are best-selling, I am not being redundant; Eagleman has written other best-selling books prior to this recent best-selling one.
54 Eagleman 2011.
55 By *work*, I mean the operational process or material means of achieving certain activities, tasks, behaviors, or other phenomena traditionally recognized as mental. See Solymosi 2012a, and Solymosi 2012b.
56 By which he means nothing more than breaking experience down into its constituent parts like neural firings and brain chemistry—not the philosophical term of art.

57 Eagleman 2011, p. 20.

58 *Ibid.*, p. 21.

59 In the closing pages of the book, Eagleman seems to shift his position by advocating neurofeedback therapy. Thanks to real-time brain scans, a person can learn to control certain brain patterns, such as the pattern of resisting the temptation to have chocolate cake. While being shown pictures of chocolate cake, the patient is also shown a pattern that correlates with his/her wanting to eat the cake and a pattern of his/her not wanting to eat it (viz. resisting). The neurofeedback pattern is a means of correlating one's phenomenology with what is going on neurally. The argument that Eagleman endorses is that resisting temptation is like working a muscle, the more you use it, the stronger it is. So by using the feedback pattern to challenge oneself to hold the neural pattern of resistance, one increases one's ability to resist later the chocolate cake. This modification of habits, as classical pragmatists would call it, is amplified by our contemporary understanding of neuroplasticity and Hebbian learning ("nerves that fire together, wire together"). Eagleman does not make the connection between habits and belief, but he leaves room for it. However, he does not seem to make the connection between resisting temptation and willpower, which is unfortunate given his philosophically-naïve dismissal of free will.

60 Note that saying science is taking a third-person perspective does nothing for Eagleman's case here. The question can be restated like this: "From my perspective, what is it like to experience from your perspective?" This only begs the question, for a scientific realist, because we are still stuck in the phenomenology of the scientist looking at the experiencer. Such an approach inevitably leads to losing ourselves in an infinite regress. The neuropragmatist, however, recognizes that experience is inherently social, that the personal grows out of the public, and that science as a public activity isn't interested in uncovering the reality behind or independent from the knower. Recall that Eagleman talks about our being "such poor observers of our own experiences." This way of phrasing—observing experience as opposed to having an experience or, simply, experiencing—presumes the validity of the veil of ideas. What I observe—on this misguided view—isn't the world itself but sense data; what the connection is between sense data and the world cannot be known by introspection yet somehow I as scientist can know that world as it is untouched by the knowing process, which, mind you, is heavily dependent upon human-made technologies and artificially constructed situations for inquiry "into" the brain.

61 A recent critique of neuroenthusiasm, by Satel and Lilienfeld (2013), aims at striking a balance between good science and good interpretation of it. While their caution is appreciated, I find their conclusions to be more on the eliminative idealist side of things. In short, they too miss the opportunity for reconstruction.

References

Baggini, Julian (2012), "Discussion/Interview with Patricia Churchland," *The Philosophers' Magazine* 57(2): 61–70.

Churchland, Patricia S. (1986), *Neurophilosophy: Toward a Unified Science of the Mind-Brain*. Cambridge, MA: MIT Press.

—(2011), *Braintrust: What Neuroscience Tells Us About Morality*. Princeton, NJ: Princeton University Press.

—(2013) *Touching a Nerve: The Self as Brain*. New York: W. W. Norton & Company.

Dalton, Thomas C. (2002), *Becoming John Dewey: Dilemmas of a Philosopher and a Naturalist*. Indianapolis and Bloomington, IN: Indiana University Press.

Dennett, Daniel C. (1991), *Consciousness Explained*. Boston, MA: Little, Brown.

—(2008), "How to Protect Human Dignity from Science," in *Human Dignity and Bioethics: Essays Commissioned by the President's Council on Bioethics*. Available at http://bioethicsprint.bioethics.gov/reports/human_dignity/chapter3.html

Dewey, John (1910), "The Influence of Darwinism on Philosophy," in *The Middle Works of John Dewey, Volume 4*, ed. Jo Ann Boydston. Carbondale, IL: Southern Illinois University Press, (Reprinted 1988).

—(1925), *The Later Works of John Dewey, 1925–1953, Volume 1: Experience and Nature*, ed. Jo Ann Boydston. Carbondale, IL: Southern Illinois University Press, (Reprinted 1981).

—(1929), *The Later Works of John Dewey, 1925–1953, Volume 4: The Quest for Certainty*, ed. Jo Ann Boydston. Carbondale, IL: Southern Illinois University Press, (Reprinted 1988).

—(1938), *The Later Works of John Dewey, 1925–1953, Volume 12: Logic: The Theory of Inquiry*, ed. Jo Ann Boydston. Carbondale, IL: Southern Illinois University Press, (Reprinted 1981).

—(1939), *The Later Works of John Dewey, 1925–1953, Volume 14: Creative Democracy - The Task Before Us*, ed. Jo Ann Boydston. Carbondale, IL: Southern Illinois University Press, (Reprinted 1988).

—(1946), "Introduction—Reconstruction as Seen Twenty-Five Years Later," in *Reconstruction in Philosophy*, by John Dewey, in the *Middle Works of John Dewey, Volume 12*, ed. Jo Ann Boydston. Carbondale, IL: Southern Illinois University Press, (Reprinted 1988).

—(2012), *Unmodern Philosophy and Modern Philosophy*. Carbondale, IL: Southern Illinois University Press.

Eagleman, David (2011), *Incognito: The Secret Lives of the Brain*. New York: Pantheon Books.

Frank, Lone (2009), *Mindfield: How Brain Science Is Changing Our World*. Oxford: Oneworld.

Hickman, Larry (2001), *Philosophical Tools for Technological Culture: Putting Pragmatism to Work*. Bloomington, IN: Indiana University Press.

—(2007), *Pragmatism as Post-Postmodernism: Lessons from John Dewey*. New York: Fordham University Press.
Humphrey, Nicholas (2011), *Soul Dust: The Magic of Consciousness*. Princeton, NJ: Princeton University Press.
Janack, Marianne (2012), *What We Mean by Experience*. Stanford, CA: Stanford University Press.
Koopman, Colin (2007), "Language is a Form of Experience: Reconciling Classical Pragmatism and Neopragmatism," *Transactions of Charles S. Peirce Society*, 43(4): 694–727.
Krikorian, Yervant H., ed. (1944), *Naturalism and the Human Spirit*. New York: Columbia University Press.
Lende, Daniel H., and Downey, Greg eds (2012), *The Encultured Brain: An Introduction to Neuroanthropology*. Cambridge, MA: MIT Press.
Lynch, Zack (with Byron Laursen) (2010), *The Neuro Revolution: How Brain Science Is Changing Our World*. New York: St. Martin's Griffin.
Misak, Cheryl (2012), *The American Pragmatists*. New York: Oxford University Press.
Moreno, Jonathan D. (2012), *Mind Wars: Brain Science and the Military in the 21st Century*. New York: Bellevue Literary Press.
Peirce, Charles S. "The Fixation of Belief," *Popular Science Monthly*, 12 November 1877, pp. 1–55.
Putnam, Hilary (2002), *The Collapse of the Fact/Value Dichotomy and Other Essays*. Cambridge, MA: Harvard University Press.
Racine, Eric (2010), *Pragmatic Neuroethics: Improving Treatment and Understanding of the Mind-Brain*. Cambridge, MA: MIT Press.
Reisch, George A. (2005), *How the Cold War Transformed Philosophy of Science: To the Icy Slopes of Logic*. New York: Cambridge University Press.
Rorty, Richard M. (1970), "In Defense of Eliminative Materialism," *The Review of Metaphysics* 24(1): 112–21.
—(1982), *Consequences of Pragmatism*. Minneapolis, MN: University of Minneapolis Press.
Satel, Sally and Lilienfeld, Scott O. (2013), *Brainwashed: The Seductive Appeal of Mindless Neuroscience*. New York: Basic Books.
Schulkin, Jay (2009), *Cognitive Adaptation: A Pragmatist Perspective*. Cambridge and New York: Cambridge University Press.
Sellars, Wilfrid (1956), "Empiricism and the Philosophy of Mind," in *Empiricism and the Philosophy of Mind*, ed. Robert Brandom. Cambridge, MA: Harvard University Press, 1997, p. 63
—(1963), "Philosophy and the Scientific Image of Man," in *Science, Perception and Reality*, by Wilfrid Sellars. Atascadero, CA: Ridgeview Publishing Company.
Snow, C. P. (1959), *The Two Cultures*. New York: Cambridge University Press.
Solymosi, Tibor (2011a), "A Reconstruction of Freedom in the Age of Neuroscience: A View from Neuropragmatism," *Contemporary Pragmatism* 8(1): 153–71.

—(2011b), "Neuropragmatism, Old and New," *Phenomenology and the Cognitive Sciences* 10(3): 347–68.

—(2012a), "Can the Two Cultures Reconcile? Reconstruction and Neuropragmatism," in *The Handbook of Neurosociology*, Jonathan Turner and David Franks (eds). Dordrecht: Springer, pp. 83–98.

—(2012b), "Pragmatism, Inquiry, and Design: A Dynamic Approach," in *Origin(s) of Design in Nature: A Fresh, Interdisciplinary Look at How Design Emerges in Complex Systems, Especially Life*, Richard Gordon, Liz Stillwaggon Swan, and Joseph Seckbach (eds). Dordrecht: Springer, pp. 143–60.

—(2013a), "Neuropragmatism on the Origins of Conscious Minding," in *Origins of Mind in Nature*, ed. Liz Stillwaggon Swan. Dordrecht: Springer, pp. 273–87.

—(2013b), "Against Representation: A Brief Introduction to Cultural Affordances," *Human Affairs* 23(4): 594–605.

Solymosi, Tibor, and Shook, John (2013a), "Neuropragmatism: A Neurophilosophical Manifesto," *European Journal of Pragmatism and American Philosophy*, 5(1): 212–34.

—(2013b), "Neuropragmatism and the Culture of Inquiry: Moving Beyond Creeping Cartesianism," *Intellectica*, 2013/2, 60: 137–59.

Tschaepe, Mark Dietrich (2009), "Pragmatics and Pragmatic Considerations in Explanation," *Contemporary Pragmatism*, 6(2): 25–44.

—(2011), "John Dewey's Conception of Scientific Explanation: Moving Philosophers of Science Past the Realism-Antirealism Debate," *Contemporary Pragmatism*, 8(2): 187–203.

6

Neuropragmatic Reconstruction

A Case from Neuroeconomics

Mark Tschaepe

Neuroscientific discourse becomes a part of our common linguistic experience after it has expanded from the realm of specialized knowledge into the realm of popular knowledge. The philosopher, John Dewey, referred to this type of expansion as the transition of data and ideas from secondary experience into primary experience. Dewey advocated applying a philosophical process, which he referred to as *reconstruction*, to the discursive knowledge that had become, or was in the process of becoming, part of our primary—or lived—experience. With regard to neuroscientific knowledge, I advocate neuropragmatic reconstruction—a subset of philosophical reconstruction—which helps sustain a fallibilist perspective toward neuroscience, thus aiding in the prevention of an all-too-blind acceptance of often exaggerated claims that are made within the realm of popular scientific writing, which have been based upon specialized, peer-reviewed scientific writing.[1]

As an example of the neuropragmatic process of reconstruction and its application to current popularized neuroscience, I provide a critical analysis of the recent work done by the neuroeconomist, Paul J. Zak, regarding oxytocin, a neuroactive hormone, and the feeling of trust. My purpose in critically analyzing this work is to reveal the increasingly strong claims made by Zak as his publications have progressively moved from the esoteric publications of neuroscience into the publications of the more popular press. In addition, my critical analysis touches upon the reviews and popular editorials that have been based upon such work, thus pushing the stronger claims of Zak's research into the public domain without providing the narrower foundation of the research upon which those claims have been based. My purpose is not to undermine Zak's research, which I find commendable. By reconstructing Zak's work, we find that the broad-brush causal claims made in the popular literature, which gradually become part of our primary experience, are a vast departure from the much narrower claims made within more specialized publications that constitute the critical secondary experience that is a largely forgotten or neglected basis of that primary experience. The reconstruction of Zak's work provides an example of the movement of neuroscientific discourse from secondary experience to primary experience.

Neuropragmatism provides a bridge by which to retrace our way back from primary experience, which has become laden with scientific language and causal claims, in order to better reassess and re-evaluate that language and those claims. In the following, I will provide the example of neuropragmatic reconstruction applied to the neuroeconomic work that has been done with regard to oxytocin and trust. This example is a means by which to reconstruct a bridge between the primary experience in which we now live and the secondary experience of scientific research upon which part of our lived experience is now based. For the sake of my critical analysis, I first offer a brief summary of John Dewey's concepts of primary and secondary experience, along with his concepts of analysis, synthesis, and reconstruction. Second, I summarize Zak's work, focusing primarily upon the trajectory from his early, narrowly focused work on oxytocin, to his later, broader claims concerning the work. Finally, I combine the process of neuropragmatic reconstruction with my summary of Zak's work, thus providing a critical picture of the transition from specialized neuroscientific work to neuroscience for popular consumption.

Primary and secondary experience

For John Dewey, scientific research entails operations for solving problems. Such research is born from *transactions* with the world of which we are a part. These transactions are what constitute experience. Most of our transactions with the world are unproblematic. It is only when problems—perturbations in the usually calm landscape of experience—do occur that we require scientific inquiry into those problems in order to solve them so that experience will again be unproblematic. *Inquiry* is an activity by which initial experience—that which is uncontrolled and indeterminate—is transformed via the processes of *analysis* and *synthesis*. Analysis is the process of discrimination by which experience is discriminated or divided into particular objects *of* experience. Synthesis is the process of identification wherein the objects of experience are unified as a whole.[2] Inquiry begins with observation of the facts at hand. Scientists discern objects of knowledge through the process of inquiry; "objects of knowledge are not given to us defined, classified, and labeled, ready for labels and pigeonholes."[3] In order to discern objects of knowledge with which scientists deal in inquiry, experience is analyzed as *data*—discernible material of experience that is divided into distinct units so that how or why a problem has taken place might be understood. In the process of synthesis, researchers utilize this data to formulate what Dewey calls *ideas*, which are suggestions or possible solutions that are to be used to solve the problem detected within experience.[4] Between the analysis of experience into data and the synthesis of data into ideas, scientists search for the location of the problem by altering experience in order to test what effects are derived from what causes.

The alteration of experience through inquiry brings about a type of experience distinct from that of "primary" experience: what Dewey refers to as *secondary* or *reflective* experience. Secondary experience is experience that has been systematically analyzed and synthesized, by which scientists attempt to gain understanding of primary experience. Through the analytic function of secondary experience, scientific objects are applied to the perturbations within primary experience; scientific objects help scientists *explain* primary experience.[5] In turn, the explanations generated via secondary experience are tested by whether or not they are warranted with regard to the ability of scientists, followed by the general public, to navigate through situations within primary experience. In the reconstruction of primary experience into secondary experience, scientists engage in the process of locating and demarcating causes of the effects they are attempting to explain.

According to Dewey, the search for "efficient causes" is the aim of scientific practice; this is always "a search for those relations upon which the *occurrence* of real qualities and values depends, by means of which we can regulate their occurrence."[6] By varying the conditions upon which various experiences occur, i.e. through the process of intervention, scientists attempt to locate the perturbation of experience so that they can formulate the cause of the perturbation, as well as possible solutions or corrections to that perturbation. In order to explain experience, scientists must submit their experience to experimentation. As Dewey claims, students must "*Produce* the cases."[7] Through the repeated and varied production of cases, combined with a fallibilist perspective toward preconceived notions of experience, scientists engage in what Dewey refers to as the process of *philosophical reconstruction*. Philosophers engage in this same process through the analysis of those cases, and the subsequent findings, that scientists have constructed.

Philosophical reconstruction is the application of experimentation within inquiry. Scientists and philosophers *re*-construct when they regard their habitual approaches to experience—their past knowledge, beliefs, and assumptions that collectively predispose them to act in specific ways—as fallible. Reconstruction provides a critical perspective toward primary experience and the tools that we have derived from analysis and synthesis within the secondary experiences of scientific research. This is the perspective I provide regarding the recent research and subsequent claims made concerning oxytocin and trusting behavior.

In the following section I provide a diachronic summary of the work done by the neuroeconomist, Paul J. Zak, as well as review and editorial work that has been based upon Zak's work. Such a summary lays the foundation for the philosophical reconstruction that is a fundamental part of neuropragmatism.

Oxytocin and its effects

Neuroeconomics is a fairly new field of study that integrates the domains of neuroscience and economics. Paul J. Zak, economist turned neuroeconomist, defines the field as one "that uses neuroscientific measurement techniques to identify neural substrates associated with economic decisions."[8] A major goal within the neuroeconomic research program, according to Zak, is "to discover proximate causes of choice behavior."[9] Most of Zak's own neuroeconomic work has been focused on an alleged connection between the neuroactive hormone, oxytocin, and the feeling of trust one person has with regard to another.

Oxytocin had traditionally been regarded as a hormone that pertained solely to lactation and childbirth in females. Its presence in males was considered irrelevant to behavior. Based upon the initial work of Lowell Getz, C. Sue Carter began to investigate a possible affiliation between oxytocin and pair bonding within mammals.[10] Since Carter began this work over 30 years ago, neuroscientific research concerning mate selection and attachment has increased, providing a wealth of data that helps address the question of why some mammals mate with the same partner and sustain that relationship beyond the act of mating, thus forming what we might refer to as "monogamous relationships."[11]

Based upon the behavioral research that had addressed pair bonding, as well as other social behaviors, Zak hypothesized an association between oxytocin (OT) and trusting behaviors in humans. The initial hypothesis put forth by Zak concerned the connection between trust, monetary transfer, and OT. As Zak states: "Specifically, we predict that when people receive a monetary transfer that is voluntary and intentional, connoting trust, peripheral OT will be higher than when people receive a monetary transfer absent an intention of trust."[12]

In order to test that hypothesis, Zak and his team collected data using a "trust game" in which participants are designated as decision-makers that are either decision-maker one (DM1) or decision-maker two (DM2). Each DM1 is paired with a DM2. None of the DMs know with whom they have been paired. DM1 is given the opportunity to transfer money to DM2 that has been given by the experimenter. The money that has been transferred to DM2 is then tripled, and DM2 can decide to return all, some, or none of the money to DM1. "This transfer is considered an index of trust because the money sent to DM2 entails a cost to DM1 which can only be recouped if DM2 voluntarily reciprocates."[13] During the study, OT measurements were taken via blood draws both before and following the decision of each participant. The research team also conducted a survey previous to the game that addressed social and developmental history, as well as affective intensity. The purpose of the survey was "to examine whether personality traits and life events were related to hormone levels and behavior in the trust game."[14]

What the research team found was that OT levels, on average, rose in each individual of the DM2 group after having received the transfer of money from an individual in the DM1 group, i.e. after a signal of trust. These different levels of OT, found before and after the exchange, were not found in individuals in the DM1 group, leading the researchers to claim: "This suggests that OT responds to signals of trust but is not associated with producing trust itself."[15] From this particular study, Zak and his team indicate that their research is an inchoate step in attempting to understand "the role of hormones in complex human social interactions that involve trust and trust-worthiness."[16] The strongest claim that they make within this particular study is that the data they have gathered leads them to "infer a causal relationship between the perceived intention behind a monetary transfer and the level of peripheral OT" and admitting that the causal direction (either an increase in OT leads to trust, or trust leads to an increase in OT) is unknown.[17] This study was published in *Hormones and Behavior* in 2005, and it was one of Zak's first scholarly publications on the subject.

A much shorter piece had been published the year before, which was claimed by the writers to have been "the first report that endogenous oxytocin in humans is related to social behaviors."[18] This study was published in the *Annals of the New York Academy of Sciences*, and the claims made were even weaker than the publication listed above. In this initial study, the strongest claim made was that "peripheral OT responds to the receipt of a social signal of trust and is statistically related to trustworthy behavior. When the social signal of trust is removed, so are the OT response and the high degree of trustworthiness."[19] From this first study, published in 2004, through subsequent papers and chapters, published through 2012, the causal language used to describe the relationship between OT and trusting behavior becomes substantially stronger.

The strength of causal claims correlates with added variables and tests, which include later incorporation of hormones such as testosterone, as well as the popularity of the press or publication in which the paper or chapter is published. As noted, the first study simply indicates a *statistical relation*, whereas the second study claims an *association*. In 2005 alone, the causal language shifts from *association* (Zak et al. in *Hormones and Behavior*), to *increases* (Kosfeld et al. in *Nature*), and finally to *facilitates* (Zak in *Behavioral and Brain Sciences*). With each subsequent publication, a stronger causal claim is made regarding the effect of OT upon trusting behavior. On the cover of *Nature*, the magazine that contains the Kosfeld et al. article, the headline goes so far as to exclaim: "The 'Trust Hormone': Oxytocin makes you believe." So what is it that changes in each study that might legitimize these linguistic shifts?

Between the 2004 study (*statistical relation*), the *Hormones and Behavior* paper (*association*), and the *Behavioral and Brain Sciences* commentary (*facilitates*), there are no differences in the research performed or the findings. There is a substantial difference between these studies and those published in *Nature* (*increases*), in which Kosfeld was the lead author. The findings in *Nature* are based upon a double-blind study that was used to "compare trusting behavior in a group of subjects that received a single dose of intranasal oxytocin with that of subjects in a control group that received placebo."[20] Whereas the previous studies are attempts to analyze endogenous levels of OT and its effects upon trusting behavior, this study is an attempt to analyze exogenous levels of OT in correlation to trusting behavior. In this study, investors were given 12 monetary units (MU) that could be transferred in increments of 4 (0, 4, 8, or 12 MU) to trustees. The amount of MU transferred was then tripled, and the trustees had the option of transferring as much MU as they desired, or none at all, back to the investor. Similar to the Zak studies, neither the investors, nor the trustees, knew who the others were, and there was no communication between the two individuals. Those investors that were given intranasal OT before the exchange exhibited a greater amount of trust in terms of transferring more money to the trustee. In order to counter the claim that OT might merely contribute to riskier behavior rather than trusting behavior, a risk experiment was also conducted that implemented a random mechanism that "'replicated the trustees' decisions in the trust experiment,' thus removing the need for any kind of trust between two parties while sustaining the element of risk for the investor."[21] Only those investors in the trust experiment that received intranasal OT exhibited any difference in the amount of money that they transferred. Trustees given oxytocin showed no differences in trusting behavior.[22] Kosfeld et al. concluded that exogenous administration of OT "causes a substantial increase in trusting behavior."[23] Such behavior, according to the authors, does not include reciprocity, as indicated by the lack of difference in behavior between the trustees administered OT and those who received a placebo. Thus, between the studies led by Zak and those of Kosfeld, the administration of exogenous OT provided the authors with what they believed was enough evidence to alter the description of the relationship between OT and trusting behavior from an *association* to *increase*, i.e. correlative claims concerning OT became causal claims.

This shift in language was sustained in 2007 when Zak et al. published a new study in *Public Library of Science ONE*: "Oxytocin Increases Generosity in Humans." The causal language remains, but the behavior described has changed from that of *trust* to *generosity*. The hypothesis upon which this study was based was that human males who were infused with 40 IU OT would be more likely to engage in what was defined as a generous behavior than those infused with a placebo. Zak and his team defined generosity as "offering more to another than he or she expects or needs."[24] They tested the amount of generosity displayed by participants by setting up two different decision tasks: the ultimatum game (UG) and the dictator game (DG). UG divides the participants in the same manner as Zak's previous experiments: dyads are formed of DM1 and DM2. In UG, DM1 is given $10 and has the opportunity to split the money with DM2, who has not been given any money from the experimenters. DM2 was then given the opportunity to reject the offer given by DM1, but if the offer was rejected, then neither party received any money. DM2 privately informed the experimenters of the minimum amount that he would accept. The potential rejection provided a way for DM2 to punish DM1 for not being generous. The researchers defined "a generous transfer in the UG as a DM1 offer that exceeds the average minimum acceptable offer."[25] The second decision task, DG, is set up in almost exactly the same manner, except DM2 cannot reject the offer made by DM1. According to the experimenters, DG provides the ability to remove generosity from the task while retaining the altruistic component.

Zak et al. found that the OT group of DM1 offered an average of 21 percent more to DM2 in the UG than those of the placebo group, while the minimum acceptable offer was unchanged by OT administration in DM2. The experimenters concluded, with regard to DM1, "that generosity was 80 percent higher in the OT group than the placebo group."[26] The administration of OT provided no substantial difference when tested in the DG. With this new 2007 study, the causal claims concerning OT had expanded to include both trusting behavior *and* generous behavior, although this particular study only pertained directly to generosity.

Zak continued to claim the causal powers of OT with regard to both trust and generosity in separate studies through 2008. In June of that year, Zak published "The Neurobiology of Trust" in *Scientific American*. This piece represented the strongest causal claims thus far, as well as the most popular publication in which these claims had been made. The cover of *Scientific American* proclaimed: "Trust Hormone: Neurobiology Reveals What Makes Us Connect." The article constitutes a general summary of Zak's work on OT. Zak begins the article by stating that he and his colleagues have demonstrated that OT "plays a major role" in the process of trust.[27] This claim, which is not necessarily causal, is followed by a brief history of Zak's work, including summaries of the above studies. The article also provides a summary of Zak's work on distrust, which has correlated high levels of testosterone—the hormone that counters the effects of OT—with low levels of trust. By the end of the article, the causal language has again shifted, with Zak claiming that "we will better understand how this simple peptide [OT] allows people to have empathy for and sustain trust in those around them, even complete strangers."[28] Although Zak refers to the complexities of environmental and internal factors to such behavior, his causal claim is still strong, having moved from *playing a major role* to *allowing* within the same article, but without providing any new findings to the research in which the terms *association*, *facilitates*, and *increases* were used.

The strength of Zak's causal claims concerning OT remains consistent through 2012. In 2012, Zak published his book, *The Moral Molecule: The Source of Love and Prosperity*. The book was published through Dutton, a popular press, thus reaching his widest audience thus far. Here OT has been granted the role of *first mover*. It is the *source*, not only of *trust* and *generosity*, but also of love and prosperity. In the Introduction, Zak claims: "My research had demonstrated that this chemical messenger both in the brain and in the blood is, in fact, the key to moral behavior. Not just in our intimate relationships, but also in our business dealings, in politics, in society at large."[29] Publishing the book led to reviews by other writers, such as Tim Harford, who wrote an article for *Financial Times Magazine*, entitled "Stock market molecules." At one point in the review Harford claims that "it is intriguing to know that the trust part can to some extent be bottled:" referring to OT.[30] In *Forbes Magazine*, Josh Bersin utilized Zak's work to help explain the importance of "empowerment" and trust in the workplace. With the re-writing of the studies of OT and behavior, these popular publications signify the transition from secondary experience to primary experience. The specifics of the scientific research not only risk being lost, but are in effect masked by the strong causal claims made within popular publications.

The causal claims made by Zak and others with regard to the research done concerning OT and human behavior by no means discount that research. What the retracing of the increasing strength of such claims within the research has done is call into question why these linguistic shifts have occurred, thereby slowing the transition from secondary experience to primary experience. This retracing of studies allows for what John Dewey referred to as philosophical reconstruction, which will provide a critical perspective to what has already been largely accepted in the popular press. Specifically, applying philosophical reconstruction to practical work in neuroscience is here referred to as *neuropragmatic reconstruction.*

Neuropragmatic reconstruction

When we reconstruct a series of neuroscientific studies, such as those of Zak, what we are doing is deliberately and creatively reviewing what has been done thus far in order to discern what possible problems arise from those studies in a definite manner instead of a manner that is "merely vaguely emotional."[31] This process is creative because we apply our critical tools to what is becoming part of our primary experience. Reconstruction is in direct contrast to merely accepting or being "submissive to what are termed authorities in different fields" without attempting to understand and question the transition from secondary to primary experience.[32] The following, then, constitutes a critical analysis of the claims made concerning oxytocin and its effects, not simply an emotional censuring of those claims. My purpose here is not to undermine the work that has been done on trust, generosity, and other social behaviors that might be associated with oxytocin. Rather, as a philosopher, I offer a critical perspective that holds the rapid advancement of claims in check, attempting to curtail exaggerated claims from becoming part of primary experience without warrant.

Before entering into a philosophical reconstruction of Zak's work that I have covered thus far, it is important to note that Zak engages in what Dewey referred to as *philosophical reconstruction* within science. Neuroeconomics directly addresses the underlying assumptions that have driven much of modern-day economics, especially the prevailing concept of *economic man* (*Homo economicus*), who allegedly bases all of his decisions upon self-interest.[33] Hypothesizing that oxytocin causes trust, which leads us to make certain economic decisions, is a direct challenge to the authority of established economic theory. Zak and other neuroeconomists have "produced the cases" that Dewey advocates, reconstructing the problem of why we behave in a trusting way by controlling and altering particular variables, thus inquiring in order to attempt to solve the question of *why* we trust or are generous when it comes to economic decisions.

Just as Zak's reconstruction of economics calls into question that which has become part of primary experience (e.g. the assumption that all rational human persons act out of self-interest), reconstructing Zak's own work calls into question the alternative he provides. At the moment of writing this, Zak's work still remains largely in the domain to which Dewey referred as secondary experience, but, given exposure of such work in the popular press, neuroeconomic ideas are shifting into the domain of primary experience, becoming part of the vulgate, just as evolutionary, genetic, and psychopharmaceutical terminology have become part of our common speech and practices. Upon analysis, the following are problematic areas within Zak's discourse that are perhaps not readily detected, especially if we were to look at only the most recent popular work without retracing the progression of that work.

First, as has been indicated throughout, is the use of increasingly strong causal claims that have been used to explain behavior. No criteria for such claims have been given explicitly, so we are left with little indication of what determines a strictly correlative connection being made between oxytocin and behavior, and what determines a causal connection from oxytocin to behavioral effects.[34] This is not a problem restricted to this particular work. Within the sciences, especially within the biological and psychological sciences, causal claims are often made without giving criteria for determining a relationship as a causal relationship. One of the tasks of neuropragmatism is to indicate this underlying assumption has been made so that we might seek to construct warranted criteria for such claims.

In direct relation to the first issue raised, we are given no indication of when research merits the use of one relationship (e.g. *association*), and when that same research merits the use of a stronger relationship (e.g. *allows*). In these cases, perhaps neuroeconomics is more forgiving with a loose array of language, but neuropragmatism, as a form of philosophical analysis that sets its sights upon neuroscientific practices, aids in detecting and questioning the effects of the transitions from one term designating a relationship to another term. Within this particular neuroeconomic project, there are significant differences in value between *statistical relation*, *association*, *facilitation*, *increase*, *allowance*, *cause*, and *source*. These differences, along with reasons for transitioning from one term to another, are never given within the literature. It is thus the task of the neuropragmatist to critically assess these studies in order to attempt to determine the meanings of these terms, as well as to attempt to determine what, if any, warrant there is for using each term in each particular instance.

Just as we critically assess the relations claimed between oxytocin and behaviors, we also critically analyze the behavioral term used within the literature. The specific terms used within Zak's studies are economic terms of art that are imported into the popular, general domain of readers largely unfamiliar with economic or neuroscientific language specific to each field. Two of the most problematic terms are *trust* and *generosity*. *Trust* is used in a very specific manner within the neuroeconomic experiments summarized above. In the domain of scientific research, this is not only unproblematic, but it is necessary with regard to controlled experimentation. Generally, each experimental model defines trust as a type of confidence or hope held by a person, i.e. an investor, that an investment will be returned in some fashion that the investor believes is fair. The way that trusting behavior is gauged within the experiments is particular to the experimental model used. The definition of trust being used does not include all trusting behavior in every possible economic situation. Rather, the operational definition used is necessarily narrow in focus so that it only pertains to the experimental model being implemented.

A sign of trust, which Zak claims triggers an increase of oxytocin, is defined as a form of *empathy*. Empathy is defined by the researchers as "an identification with and understanding of another's situation or feelings."[35] In a 2011 study concerning the effects of empathy upon oxytocin release and subsequent generosity, the specific empathic model used was the viewing of an emotional video with which the viewer could identify. After viewing the film, some subjects were asked to participate in the UG task. Those who had an increase in experienced empathy were found to be more generous in the UG than the control group.[36] The definition of generosity used is "offering more to another than he or she expects or needs."[37] The claim being made with regard to the behaviors associated by experimenters with these terms is: a sign of empathy leads to an increase in oxytocin levels, which leads to an increase in trusting behavior or generosity. As indicated, these terms have specific meanings that have been operationalized for the sake of the experiments.

The experiments performed to test the effects of empathy, increases in oxytocin (as well as variations in other hormones, such as decreases in testosterone and increases in cortisol), and increases in trusting behavior and generosity, do not change when the findings from the experiments are published in the popular press. What does change, as indicated above, is the strength of the causal claims that are made. What also changes is the detail given concerning the experiments. When strong causal claims are combined with the use of common terminology that has been specifically operationalized for the sake of experiments, the probability of the audience interpreting that terminology as being used with less specific, more generalized, meanings, likely increases. The use of general claims contributes to this interpretation. For instance, Zak states at the end of his book, "Oxytocin—a reproductive hormone—makes us moral, so ultimately, you could say that we are moral because of our origins as sexual creatures."[38] This is a far cry from the 2004 statement that "peripheral OT responds to the receipt of a social signal of trust and is statistically related to trustworthy behavior."[39] The amplification of the 2012 claim is made readily apparent through the process of reconstruction.

Through the above reconstruction of the work that Zak and other neuroeconomists have done with regard to oxytocin and behavior, we are left in a position wherein we can accept the findings of their experiments without either accepting all of their claims wholesale, or rejecting all of their work. The neuropragmatic critical assessment of the work they have done leads us away from blind acceptance and toward the position of being informed individuals. This is the type of position to which Dewey referred, which entails throwing "off an outer slavery to second-hand and ready-made opinions."[40] From this position we are better able to inquire into how much of the secondary experience provided by the work of neuroeconomists "is validated and verified in present need, opportunity, and application."[41] In addition, we are equipped with a set of tools from the reconstruction of this work that can be applied to future studies and proposals that integrate neuroscience, economics, and morality. Regarding the work reviewed above, we need not accept that oxytocin is the cause of all trust and generosity, but it does seem an important tool with which to understand those behaviors, and Zak has provided specific ways to understand its relation to those behaviors. Reconstructing the work that has provided us with this tool does not answer why we trust or are generous, but it does allow us to narrow that question in ways that allow for specific answers within specific contexts. Re-evaluating the research concerning oxytocin and behavior re-enforces its value while curtailing its exaggeration.

Conclusion

Neuropragmatism involves the active reconstruction of neuroscientific research, especially research that is becoming incorporated into our primary experience. In the case examined here, there has only been an eight-year time span between the initial publication correlating oxytocin and trust, and Zak's book being released in the popular press. One of the roles of the neuropragmatist is to critically assess the data and the subsequent ideas that have become part of the secondary experience of neuroscience, which are being implemented in order to identify and understand the problems within primary experience. The neuropragmatist does not reject the work of the neuroscientists, but rather assists on the bridge between the secondary experience of neuroscience and the primary experience of the non-neuroscientist.

Scientists, in this case, neuroeconomists, analyze primary experience into particularized objects—data—that are the tools of experimentation, such as percentages of peripheral oxytocin present at a given time within a subject. The results of the experiments are then synthesized into ideas, such as the idea that increases of oxytocin increase generosity. These ideas are initially utilized as secondary experience—scientific objects of knowledge—that is applied to primary experience by scientists. With familiarity, often facilitated by the translation of scientific research into the popular press, the scientific objects of knowledge become part of our primary experience. Often, the specific details of terminology become lost in translation. In addition, ungrounded causal claims may become stronger, just as they have in the work discussed here, as ideas are transmitted through more readily available media.

The process of neuropragmatic reconstruction is a critical reflection that retraces the construction of neuroscientific discourse in order to assess the possible problems that emerge from the expansion of that discourse as it moves away from specialized domains, into more general domains. In a similar manner as genetic, evolutionary, and psychopharmaceutical discourse, neuroscientific discourse is rapidly becoming part of our primary experience. One task of the neuropragmatist is to act as a liaison officer who assists in the transport of neuroscientific secondary experience into the realm of primary experience in an informed and responsible manner.

Notes

1 The specifics of specialized scientific discourse being lost in popular discourse are evident in many scientific branches. For example, it is not uncommon to read that a "gene for obesity" or a "gene for addiction" has been discovered, although geneticists do not actually claim to have "discovered" genes for such general traits.
2 Dewey 1961–91, *Later Works, vol. 8*, p. 275. Hereafter *LW8*.
3 Dewey 1961–91, *Later Works, vol. 1*, p. 170. Hereafter *LW1*.
4 Dewey *LW8*, pp. 197–8.
5 Dewey *LW1*, pp. 15–16.
6 Dewey 1961–91, *Later Works, vol. 4*, p. 83. Hereafter *LW4*.
7 Dewey 1961–91, *Later Works, vol. 12*, p. 275. Hereafter *LW12*.
8 Zak 2004, p. 1737.
9 *Ibid.*
10 Cf. Getz et al. 1981.
11 Cf. Aragona and Wang 2004; Carter and Getz 1993, 1998, 2005; Insel and Hulihan 1995; Wang and Aragona 2004; Winslow et al. 1993.
12 Zak et al. 2005, p. 522.
13 *Ibid.*, p. 523.
14 *Ibid.*, p. 524.
15 *Ibid.*, p. 525.
16 *Ibid.*, p. 526.
17 *Ibid.*
18 Zak et al. 2004, p. 224.
19 *Ibid.*, p. 226.
20 Kosfeld et al. 2005, p. 673.
21 *Ibid.*, p. 674.
22 *Ibid.*, p. 675.
23 *Ibid.*
24 Zak et al. 2007, p. 1.
25 *Ibid.*, p. 2.
26 *Ibid.*
27 Zak 2008, p. 88.
28 *Ibid.*, p. 95.
29 Zak 2012, p. x. Zak does clarify later in the Introduction, stating: "Obviously there's more to it, because no one chemical in the body functions all alone, and other factors from a person's life experience play a role as well." He cannot wholly give up the strength of his claim, though, going on to say: "But as we'll see in the chapters ahead, oxytocin orchestrates the kind of generous and caring behavior that every culture, everywhere in the world, endorses as the right way to live, the cooperative,

benign, pro-social way of living that every culture everywhere on the planet describes as 'moral.'" (p. xii).

30 Harford 2012.

31 Dewey 1961–91, *Middle Works, vol. 12*, p. 161.

32 Dewey 1961–91, *Later Works, vol. 5*, p. 135. Hereafter *LW5*.

33 Zak 2012, p. 21.

34 Zak mentions the transition from correlation to causation in his 2004 piece, "Neuroeconomics," which was published in *Philosophical Transactions of the Royal Society of London*, but does not state any criteria for the transition within the literature (p. 1745).

35 Zak et al. 2007, p. 2.

36 Barraza and Zak 2009, p. 188.

37 Zak et al. 2007, p. 1.

38 Zak 2012, p. 211.

39 Zak 2004, p. 226.

40 Dewey *LW5*, p. 139.

41 *Ibid.*, p. 142.

References

Aragona, Brandon J., and Wang, Zuoxin (2004), "The Prairie Vole (Microtus Ochrogaster): An animal model for behavioral neuroendocrine research in pair bonding," *Institute for Laboratory Animal Research Journal (ILAR)*, 45(1): 35–45.

Barraza, Jorge A., and Zak, Paul J. (2009), "Empathy toward Strangers Triggers Oxytocin Release and Subsequent Generosity," *Annals of the New York Academy of Science*, 1167: 182–9.

Bersin, Josh (2012), "How the Trust Hormone Drives Business Performance," *Forbes*, April 30. Available at: www.forbes.com/sites/joshbersin/2012/04/30/how-the-trust-molecule-drives-business-performance/

Carter, C. Sue (1998), "Neuroendocrine Perspectives on Social Attachment and Love," *Psychoneuroendocrinology* 23(8): 779–818.

Carter, C. Sue and Getz, Lowell L. (1993), "Monogamy and the Prairie Vole: Studies of the prairie vole—a secretive, mouselike animal—have revealed hormones that may be responsible for monogamous behavior," *Scientific American*, June: 100–6.

—(2005), "Biological Perspectives on Social Attachment and Bonding," in *Attachment and Bonding: A New Synthesis*, C. Sue Carter et al. (eds). Cambridge, MA: MIT Press, pp. 85–100.

Dewey, John (1961–91), *The Collected Works of John Dewey, 1882–1953*, ed. Jo Ann Boydston. Carbondale, IL: Southern Illinois University Press. Published as *The*

Early Works: 1882–1898 (*EW*), *The Middle Works: 1899–1924* (*MW*), and *The Later Works: 1925–1953* (*LW*).

Getz, Lowell L., Carter, C. Sue, and Gavish, L. (1981), "The Mating System of the Prairie Vole Microtus Ochrogaster: Field and laboratory evidence for pair-bonding," *Behavioral Ecology and Sociobiology* 8: 189–94.

Harford, Tim (2012), "Stock Market Molecules," *FT Magazine*, June 9. Available at: www.ft.com/cms/s/2/e8a01104-af6f-11e1-a8a7-00144feabdc0.html#axzz1zC0zQcAK

Insel, Thomas R., and Hulihan, T. J. (1995), "A Gender-Specific Mechanism for Pair Bonding: Oxytocin and Partner Preference Formation in Monogamous Voles," *Behavioral Neuroscience* 109: 782–9.

Kosfeld, Michael, Heinrichs, Markus, Zak, Paul J., Fischbacher, Urs, and Fehr, Ernst (2005), "Oxytocin Increases Trust in Humans," *Nature* 435(2): 673–6.

Wang, Zuoxin and Aragona, Brandon J. (2004), "Neurochemical Regulation of Pair Bonding in Male Prairie Voles," *Physiology & Behavior* 83: 319–28.

Winslow, J. T., Hastings, N., Carter, C. S., Harbaugh, C. R., and Insel, T. R. (1993), "A Role for Central Vasopressin in Pair Bonding in Monogamous Prairie Voles," *Nature* 365: 545–8.

Zak, Paul J. (2004), "Neuroeconomics," *Philosophical Transactions of the Royal Society of London* 359: 1737–48.

—(2005), "Trust: A temporary human attachment facilitated by oxytocin," *Behavioral and Brain Sciences* 28(3): 368–9.

—(2008), "The Neurobiology of Trust," *Scientific American*, June: 88–92.

—(2011), "The Physiology of Moral Sentiments," *Journal of Economic Behavior & Organization* 77: 53–65.

—(2012), *The Moral Molecule: The Source of Love and Prosperity.* New York: Dutton.

Zak, Paul J., Kurzban, Robert, and Matzner, William T. (2004), "The Neurobiology of Trust," *Annals of the New York Academy of Science* 1032: 224–7.

—(2005), "Oxytocin is Associated with Human Trustworthiness," *Hormones and Behavior* 48: 522–7.

Zak, Paul J., Stanton, Angela A., and Ahmadi, Sheila (2007), "Oxytocin Increases Generosity in Humans," *PLoS ONE* 2(11): e1128 (1–5).

7

The Most Important Thing Neuropragmatism Can Do

Providing an Alternative to "Cognitive" Neuroscience

Eric P. Charles, Sabrina Golonka, and Andrew D. Wilson

The long-known problem

> *The frantic search for sources of motivation and of emotion in visceral activity, though initiated by introspective analysis, has been supported by the faith that the nervous system is only a conductor having no sources of energy within itself ... [Theories derived from this approach] may or may not be true, but their truth must be demonstrated by experiment and cannot be assumed on a background of questionable neurology ... The facts of both psychology and neurology show a degree of plasticity, of organization, and of adaptation in behavior which is far beyond any present possibility of explanation. For immediate progress it is not very important that we should have a correct theory of brain activity, but it is essential that we shall not be handicapped by a false one.*
>
> Karl Lashley[1]

Karl Lashley was, by several accounts, the greatest psychologist-neurologist of the early twentieth century. He is most well known for his search for the "engram," the biological "memory trace" in the brain. He eventually abandoned this search, concluding correctly that the engram did not exist. Despite that, his search has been reinvented in modern times, with a slightly different vocabulary, in light of modern advances in neuroimaging. The researchers on this new quest seemed to think that Lashley's failure was due to technological limitations—if only that brilliant man had a 10 tesla fMRI machine, then he would have found it for sure! But that was not the problem. The problem was that the central metaphor guiding his search did not match reality in important ways; there simply is not a *place* in the brain where memories are *stored*. The biological events involved in the adaptive change of behavior over time, and even the biological events involved in re-experiencing the past, do not resemble the process of retrieving information from a reliable, representational storage system. What was once a perfectly viable hypothesis about how the mind works is simply not true.

Even though cognitive scientists now increasingly recognize that this hypothesis was false, many neuroscientists continue to use language that is derived from it. While some will protest that we are exaggerating, an examination of the literature reveals that neuroscientists continue to rely on the implications of the (false) mind-is-computer hypothesis. Why do they do this? We suggest that they continue because they have no alternative, and that the greatest service neuropragmatism can offer is to provide a better way of talking about the role of the brain.

Cognitive neuroscience—the science of a bad hypothesis

Cognitive neuroscience was known to be foolish before it began. "Cognitive psychology" in its modern form, is an aggressive use of a computer metaphor to explain how the mind operates. This approach had some important virtues. It led to many discoveries about mental phenomena, and it freed researchers from the oddly restrictive behaviorism of a prior era. However, while we might one day have machines that work like brains, our brains do not work much like current computers, and certainly not much like the computers of the late 1950s and early 1960s. While the new metaphor has been very productive, when it is pushed to the extreme it is as silly and restrictive as the reductive behaviorism it was supposed to replace. It consists, for the most part of a proliferation of "hypothetical constructs"—such as reliable "memory stores" or "intelligence"—which are conceived as inherently non-observable things-in-the-head that can be offered to explain behavior.

MacCorquodale and Meehl wrote the first paper to meaningfully distinguish hypothetical constructs from related notions.[2] They argued persuasively that there is nothing inherently wrong with hypothetical entities, and that the hard sciences use them quite effectively. However, they caution psychologists against two mistakes:

> First … there is the failure explicitly to announce the postulates concerning existential properties, so that these are introduced more or less surreptitiously

> and ad hoc as occasion demands. Secondly, by this device there is subtly achieved a transition from admissible intervening variables to inadmissible hypothetical constructs. *These hypothetical constructs*, unlike intervening variables, are inadmissible because they *require the existence of entities and the occurrence of processes which cannot be seriously believed because of other knowledge.*[3]

They later rephrase the second point more forcefully, reminding psychologists that: "It is perhaps legitimate, even now, to require of a hypothetical construct that it should not be manifestly unreal in the sense that it assumes inner events that cannot conceivably occur."[4]

Without passing judgment on the early days of cognitive psychology, we can report that the current field makes those mistakes flagrantly. It is quite likely that individual research programs are relatively consistent in their use of hypothetical constructs, but *across* programs the divergence is tremendous. As has been pointed out elsewhere, the terms of cognitive psychology give a false sense of concreteness, while really being hopelessly vague:

> For example, if you were interested in "how memory affects perceived meaning," how would you go about investigating it? "Memory" could be operationalized in a wide variety of ways; memory could refer to what people remember in the present, any change in behavior following an event you (the investigator) are sure happened, the effects of bodily (neuronal?) alterations on future behavior, the retention of a conditioned response, etc. "Perception" and "meaning" are similarly, if not more, ambiguous. Perhaps you will ask people to recall certain events in their lives, then have them interpret ink-blots; perhaps you will flash lists of words for 50 milliseconds at a time, then see if they feel positively toward words you repeated several times; perhaps you will ask people the perspective from which they view specific memories, then test them for their accuracy in recalling important aspects of the event; the variety is almost infinite. Despite this incredible lack of specificity of the question, it is easily transformed into a concrete empirical endeavor. Whatever the form of the study, the results could be reported in the local newspaper and the average high school student reading it will nod their head as if they are learning something important about human nature. Further, if you also give the task to chimps or dolphins it can make national headlines—"Chimps perception of meaning less affected by memory manipulations than teenagers"—despite the headline saying nothing concrete, it is perceived as understandable and straightforward.[5]

Cognitive neuroscience is, alas, the process of looking inside the brain to try to find the hypothetical constructs of cognitive psychology. As the quote from

Lashley at the beginning of this chapter should make clear, this process did not start with the relatively recent advent of neuroimaging, it was alive and well at the start of the twentieth century. To the extent that the terms of cognitive psychology can be made concrete, they are arbitrary; and, to the extent that the original meanings of the terms are held to tightly, they don't refer to anything that actually happens in the brain. Thus, the terms either don't mean anything obvious, or they are obviously wrong.

Many cognitive neuroscientists, to their credit, clearly recognize this problem. The amazing models of brain activity currently being offered by cognitive neuroscientists retain the terminology of cognitive psychology, while jettisoning most of the terms' obvious meanings. The importance of neural network modeling, for example, is that it provides a model for behavioral change without anywhere having stored "memories," a central "thinking" system, or a clearly identifiable "representation" of the difference-in-the-world that leads to the different responses of the system. And yet, those words are commonly used in discussions of neural networks; the terms remain, though they are stripped of any semblance of their original, intuitive, natural meaning. The terms remain and cause endless confusion among those trying to present and understand the implication of the research being conducted. The terms remain, because there is no other language to use.

Is the language really so corrupting? Yes. We did a cursory review of the *Journal of Cognitive Neuroscience, Volume 20* (2008), a journal with an impact factor to make all but the very top psychology-specific journals envious. Focusing first on memory-related papers:

- We encountered much talk of "false memory." This seemed perfectly fine at first, until we realized that in most other articles "memory" clearly indicates *accuracy* with respect to past events. In those articles, when one is mistaken about the truth of the situation, one has *failed to remember*. Thus, depending on the article, a mistake about past events is either defined as "false memory"—a variety of memory—or the lack of any variety of memory.
- We also learned that the hippocampus and perirhinal cortex contribute to "episodic item and associative encoding." While this suggests that items, and the associations between items, are somehow stored in the identified areas of the brain, it actually just means that those areas of the brain fire more frequently shortly after you see something new. It is quite clear that, despite cognitive neuroscientists' renunciation of dualism, they cannot

communicate without the notion that something is somehow stored in the brain. But the brain is not a bank vault into which things are placed for later retrieval, the brain is an organ that adapts to the environment over time.

- Distracted, we citation-jumped a bit into the future, learning that the ventral regions of the posterior parietal cortex "binds episodic features stored in disparate neocortical regions." All fine and good on the surface, until we remembered that features are *not* stored anywhere, and that the metaphor of "binding" suggests that we have reified the memories as things that can be attached to other things while some third-thing wraps around them.

Returning to *JOCN* 2008, we examined work on the "central executive," which was supposed to be a place-holder, a term for stuff we didn't understand. We found out that the thalamus is particularly crucial to activities of this non-entity.

"Processing," another common term, might seem neutral, but, in fact, use of the term assumes a very odd notion of what the brain does. For example; one article that assumed the brain was an information-processing system starts by telling us that there is "disagreement regarding the nature of the information that is maintained in linguistic short-term memory." Of course there is! That is because "information," "maintain," "linguistic," and "memory" are all highly ambiguous terms, which likely, even in their less ambiguous moments, do not reflect what the brain was doing during the intervals of interest to the study (i.e. waiting patiently in a brain scanner to see if a second stimulus matches a first).

Not all is bleak though. There were somewhat more acceptable suggestions that a given part of the brain could be "associated with," "correlated with," "coded to," "involved in," "activated during," "participating in," "underlying," "playing a critical role in," or "supporting" different psychological happenings. There was also discussion of how one area of the brain "inhibits" or "activates" another area of the brain, which is at least better than the claim that a brain area "inhibits" or "activates" a psychological happening.

The latter language is more neutral, and hence better than the cognitive language. Also on the positive side, a brief scan of more recent volumes of this journal suggest that they now favor use of the neutral references to what the brain does, whereas earlier volumes seemed to favor the egregious descriptors. However, while this is a step in the right direction, the neutral descriptors are not very satisfactory. They are about as useful as telling us that sparkplugs "participate in" the movement of a car.

What to do?

As many of the other articles in this book argue, your mind is more than the activity of your brain; your mind extends into your body, and into your environment, including the social elements thereof. However, neuroscientists interested in the mind continue to use the inaccurate, brain-focused language given to them by cognitive psychology. They can be forgiven for one obvious reason: there is no alternative. Neuroscientists want to do their work, publish it, get funding, get coverage by the press, etc. They want to study the brain. Hence, they can be forgiven for not fixing the language of psychology; they can be forgiven for stretching the terms they are given beyond any semblance of the terms' original meanings because, frankly, neuroscientists have better things to do. The psychologists, and philosophers of psychology, on the other hand, have been purely negligent in their duty. Philosophers have not only been complacent in the abuse of language, they have facilitated it willfully. For example, philosophers have worked hard to warp the definitions of "representation" until it bears no obvious resemblance to the notion of presenting-something-again, as in the way a picture of your childhood home re-presents certain aspects of the home. Psychologists as well, or at least those who have recognized the problem, have shied from their clear duty.

Which psychologists are most to blame? Presumably those most guilty are those who complain the most about the existing language. Rather than sit idly by, complaining about silly things their colleagues say, these psychologists should be providing more intelligent ways to discuss what the brain actually does. There are many groups of psychologists who fit into this category, but we can only speak in more depth about those with which we are most familiar: Ecological Psychologists and Radical Behaviorists. While the opinions represented by these fields are considered "minority" opinions in psychology, they are quite large and empirically successful minorities, generating experimental and observational research at the cutting edge of psychology, kinesiology, education, and neuroscience. Moreover, it should be noted in this context that these traditions are intellectually descended from pragmatism.[6] Thus, you might expect them to provide a good starting point for neuroscientists looking for alternatives to cognitive psychology.

Why, then, have these approaches not provided viable alternatives? Because, while it is clear that the fields of behaviorism and ecological psychology can generate alternative means for speaking about the brain, there are no individuals in the fields clearly obligated to do so. Further, due to the history of these two fields, individual researchers tend to shy away from brain work, focusing more on a behavioral level of analysis. This is not to say that brain work is not done in these fields, but to point out that existing work gives only limited hints as to better ways to talk about the brain. What is needed is a fully formed alternative language. While we cannot possibly provide that here, we can anticipate certain aspects of what will come.

Ways of speaking about the brain

> *I don't think it's meaningful to say I currently (sitting at my desk) have the ability to hit a softball; that ability isn't stored somewhere for me to access on demand. I do think it's closer to say I have the ability to become something that can hit a softball.*
>
> Andrew D. Wilson[7]

We believe there are two ready sources of insight into how we can better speak about the brain. First, there is a rich foundation of conceptual insights set up in the early 1900s by those attempting to reimagine psychology in light of, among other things, insights from pragmatic philosophy. Second, there is a rich variety of recent experimental and observational results that reveal flaws in the now-accepted vision of the psychology–neuroscience relationship, suggesting that those early intellectual risk takers were really on to something. Our presentations of both of these foundational elements must be brief, and selective, but we hope even in this limited space to offer a taste of the potential for neuropragmatism to lead us to better ways of speaking about the brain and its function.

Speculative Neuroscience, circa 1936—levels upon levels of loops and dynamics

Perceptual research under the name "Ecological Psychology" emerged out of the work of James J. Gibson in the early 1960s, and the movement has tremendously advanced our understanding of perception-action systems. It is related to some of the more molar approaches to behaviorism, and it overlaps with the perceptual-control theory that grew out of Power's work around the same time period.[8] A key lesson out of the ecological approach is that organism and environment are not connected by simple input or output relations, but rather by dynamic, circular relations. This is not a trivial observation, as circular relations allow for easy solutions to problems that would be very difficult for simple input–output mechanisms to solve. A body that is part of such a circular relation can rely on stable aspects of the world to accomplish many tasks without needing much "computation" on the part of the brain.[9]

Perhaps the most aggressive expression of this idea was made by E. B. Holt, a proto-behaviorist, proto-ecological psychologist, and proto-epigeneticist.[10] Almost eighty years ago, in a chapter dedicated to the Russian physiologist Beritoff, Holt tried to describe the multiple, embedded levels of circularity available to control behavior.[11] In "Eight Steps in Neuro-muscular Integration," Holt attempts to show how the basic principles of a sensory-motor loop can account for numerous aspects of behavior, and behavioral development. His proposed system is ambitious, and ultimately a bit unwieldy, but it shows how much a system using the organism-in-the-world approach of pragmatism can simplify what we think the brain needs to do.

Lowest level loops: Circular reflexes

In the lowest level loops, a motor neuron causes the stimulation of sensory neurons, which in turn affects the firing of the motor neuron. For example, when a neuron's firing tenses a muscle it squeezes proprioceptive cells, and activation of that proprioceptive cell will typically affect the firing of the original motor neuron, leading to muscle relaxation. Two neurons, properly connected, can thus maintain static positions or smooth forces, with no additional input required. Such loops greatly simplify, for example, the challenge of holding your posture. This suggests that, while a brain might adjust the strength of connections between neuronal loops at this level, depending on the posture-to-be-obtained, it might need to do little else to maintain such postures.

Level 2: Reciprocal innervation

As circular reflexes of antagonistic muscle groups are integrated together, neurons that flex one muscle are linked to the neurons that relax other muscles. Such loops greatly simplify, for example, the challenge of lifting and lowering limbs. This suggests that the brain need not tightly control all muscle groups involved in a movement, as the muscles are innervated to handle much of that coordination peripherally.

Level 3: Adient reflexes

To be "adient" means "to approach." Further integration of sensory-motor loops can easily explain how organisms orient-toward and approach objects and events. Without guidance from a central system, motor neurons strengthen their connection with sensory neurons until any sources of a given stimulation can produce the same movements by which an organism self-produces that stimulation. For example, a human fetus makes a lot of random movements, including finger flexes that result in palm touching. "Canalization" of that loop results in a "reflex" whereby fingers flex in response to palm stimulation, i.e. the newborn baby's grasping reflex. This same principle can be applied to other perceptual systems to help explain phenomena such as auditory imitation.

While we are here dealing with processes that, at least in higher organisms, will likely involve neural activity in the brain, it should be clear that no "top-down," homunculus-like process has been invoked. Even without that, these loops provide quite a bit of sophistication: because they will naturally develop so as to function effectively given any "special posture from which a movement starts," they can accommodate in-process perturbations equivalent to changes in initial posture, all without intervention from a "higher-function."

Level 4: Concatenation of reflexes

As Level 3 processes become chained together, we find it ever more accurate to describe what an animal is doing in terms of behavior *directed toward* some external thing. That is, while Level 1, 2, and 3 loops can be described in terms of the immediate stimulation at a sensory neuron, the integration of such systems leads the organism to be responding simultaneously to several properties of a particular object—and to respond to the conjunction of several different properties in a particular physical location *is* to be responding to the object in which those properties conjoin. In this context, Holt specifically invokes William James's notion of habits, and connects habit formation with broader discussion of the "association" of "images" and "ideas." We believe that Holt is pointing out how the strengthening of neuro-muscular integration at this level can explain behaviors ranging from a dog drooling at the sight of a particular laboratory assistant, to the tipping of a hat at the sight of an old acquaintance. The behavior in each case is controlled by a loop from the presence of a complex, multi-propertied object or situation to a particular behavior or set of behaviors, with no higher control needed.

Level 5: Movements of exploration

Here we combine loops from the previous levels and add in a bit of random movement to create the exploratory behavior one sees, for example, when a ball is placed in a toddler's hands. If these experiences repeat, then new patterns of action are set up, such that the same "exploratory" movements can happen even in the absences of the ball. (If you can hold your hands out in front of you and pretend to be touching the outside of a ball, then you know what Holt is talking about.) We now start to see the implications of perception-action loops for more clearly "psychological" processes. As Holt points out, to perform the initial exploratory movements is, literally, to *feel* the contours of the ball. To be able to perform the same motions without the ball present is then, in some sense, to remember the ball (to feel-it-again) despite its absence.[12] Holt is thus providing initial hints as to how organism–environment loops can produce the rudiments of "memory." This gets our foot in the door to higher functions, without needing to invoke a store of re-presentations, or other similar myths, and certainly without some central system that indexes, retrieves, and re-interprets what is "stored" in other parts of the brain. The loop itself, strengthened sufficiently through development, can account for the re-behavior.

Levels 6–8

Alas, at this point it becomes clear that while Holt will continue to ambitiously try to explain ever "higher" psychological functioning through higher-level loops—all without needing a central control mechanism to direct such functioning—he will not move us further toward developing a vocabulary to describe the role of the brain in such processes. Level 6, the interconnecting of multiple exploratory levels, explains, for example, how hearing a song invokes a visual "image" of past events.[13] Level 7, the strengthening and interconnecting of several Level 6 mechanisms, allows for what we might call "imagination," as such loops allow people to act in elaborate ways toward objects that are not currently present. This level also begins the appearance of "self-determination," as the triggering stimuli have little obvious connection with the behaviors themselves, as when a certain smell leads someone to quietly sing a song to themselves and perhaps even begin to recreate movements from a high-school dance. In such a situation, there *are* specific causal relations to be discovered, created by the canalization of certain neural connections due to past events, but the causes of the behaviors are "non-obvious."[14] Finally, Level 8 deals with the integration of responses to multiple objects or events, which will occur in any sufficiently complex organism. Holt admits that this might not be impressive as a final level of integration, at least at first blush. However, he believes it is key for understanding phenomena like successful navigation of the environment, and intelligent behavior in general (i.e. behaviors that must occur with respect to several simultaneous aspects about the world).

Though Holt could lay all this out in 1936, he was working largely in a vacuum, based on his own personal intellectual connections with William James. Most of his colleagues were creating the foundation for cognitive psychology, and his students had not yet developed the behavioral and ecological systems that would lead to their future empirical successes. There is now much evidence to support Holt's notion of loops from bottom to top: loops that provide flexibility and redundancy, with continuous environmental support playing an essential role all the way from bottom to top. Top researchers are currently struggling valiantly in an effort to come to grips with the role of the brain in such a system, and in so doing, a new vocabulary is slowly, and awkwardly, starting to emerge.

A modern update: What is the brain doing?

It is clearly the case that the brain is up to something important. In humans, it consumes approximately 20 percent of our body's energy[15] and our bodies are thoroughly innervated by projections to and from the nervous system. So what, precisely, is it up to?

Real brains are always busy[16] and are utterly integrated with a wide variety of other *internal* systems (e.g. musculo-skeletal, circulatory, and cardiovascular) as well as to the *external* world around it, and also to some systems that sit ambiguously between internal and external (e.g. the digestive and respiratory systems). In each of these cases, the integration of the systems is mediated via perception. Successful activity, therefore, entails perceiving and controlling both these internal and external resources to solve specific tasks; controlling both the *inherent dynamics* of the system and the *incidental dynamics* due to the situation.[17]

Brains presumably play a critical role in this process, but they aren't the only player in the game, and there are a lot of things brains simply don't *need* to do because some other system takes care of it. For example, in the "equilibrium point" model of limb control, the brain does not need to explicitly set a series of joint angles to move into a given joint configuration; instead, the nervous system merely needs to set a force-length relation for the muscles at the joint, and the limb will smoothly move to its new equilibrium point, with the specific trajectory determined completely by the local anatomical constraints.[18] One is tempted to say that the local constraints "select" the trajectory, but of course almost everything implied by the term "select" does not apply, and the more useful description is that the trajectory "emerges" from this constellation of constraints on a system with a specific dynamic. In effect, the anatomical and physiological organization *of the arm* greatly simplifies the situation, allowing the brain's role in smooth movement to be efficiently minimal.

This principle implies the following: In order to work out what the brain *does* need to do, we need a clear theory about what everything else is up to, so that we don't waste time looking for the neural implementation of something taken care of elsewhere in the body. In other words, we need a *job description* for the brain. To develop this job description, we will first lay out what the non-neural components of extended, embodied cognitive systems do to support behavior. This will then leave us with work to be done by the brain, and the resulting gap suggests a very different role than the traditional cognitive model.

Bodies as task-specific devices

Cognitive systems are not general purpose systems; we are, instead, at any given moment, one kind of *task-specific device* (TSD).[19] A TSD is a smart solution to a specific problem. *Smart* devices[20] take advantage of external, task-specific resources, as well as the internal resources described below, to find and use locally optimal solutions to problems. The kind of device we currently are is a function of what we've been up to recently; the specifics of the device reflect the nature of multiple subsystems and how these respond to the task space we are currently embedded within. These devices must therefore be *assembled* and *controlled* with respect to task-specific constraints (this requires perception of those constraints). Specifically, these TSDs must be *softly* assembled. To be "softly assembled" is to be built out of whatever resources are available to solve the task at hand, resources which can be removed and used again later as part of a different TSD. This is in contrast to a device that is built to do one particular thing, and can only ever do that thing. Take a crude example: I can use my hand to pat someone on the head or to pull their hair. The resource (the hand) is the same, but adjustments to the larger system (the body) have placed it in a different relation to the other element of the task (the head). The most likely primary role for the nervous system is going to be in supporting this soft assembly of task resources; our first task, therefore, is to determine what kind of resources we are talking about.

"External" resources (Gibsonian information)

An essential and vastly underappreciated aspect of the world is the flux of energy in the "empty space" around us, the light and pressure waves criss-crossing, chemical gradients, magnetic fields, etc. Gibson labeled the structural elements of this ambient energy that can support behavior "information."

This information flow is surprisingly stable, as is our access to it (following the requisite learning).[21] We are not adrift in a "blooming, buzzing confusion" (to use the oft misunderstood William James quote). As with other stable aspects of our ecological niche, we can (and evidence shows that we do) rely on this information to do a lot of work for us. We exist in and move through a flow of information and our behavior emerges as we interact with that flow. There is no need to construct a model of our environment; as Rodney Brooks famously claims, we can let the world be its own model.[22]

We are not passive blank slates, however. At any given moment in time we are very specific measurement devices, sensitive to some information variables but not others, and capable of responding in some ways but not others. We spend our days being one device, then another, then another, in response to changes in the information flow, and in accordance with our capacities. The information flow alters as our location in space and time changes, and is specific to the current environment.

Internal (non-neuronal) resources

The human body has various subsystems with different capacities from which to assemble the various TSDs that we can become. Each has their own dynamical properties and typically operates over different time scales. These subsystems are nonlinearly coupled to each other in a variety of ways, but can be separated out for analytical purposes. What follows develops the taxonomy described by Bingham.[23]

The *link-segment skeletal system* provides a very stable base for the other systems; the bones are all different sizes and shapes and are attached to each other with very particular joint configurations. The overall configuration does not radically alter over our lifetimes without serious injury, although it does change in response to prolonged use (e.g. thickening with weight training, thinning in extended weightlessness). The role of this system is to provide a physical substrate for the transmission of the forces involved in moving and interacting with the world. The *musculo-tendon system* is also globally quite stable in its configuration, but its form is more responsive to use on time scales ranging from days to weeks (e.g. weight-bearing exercise increases the size of muscles and stretching alters the give of tendons). The role of this system is to generate the forces required to move the link-segment system; it's organized in ways that solve many of the "degrees of freedom" problems inherent in controlling a complex system.[24] The *circulatory system* is highly responsive to current events: it only takes seconds for your heart rate to accommodate current energy demands, and veins and arteries change size and shape in response. The role of this system is to deliver energy on demand to the musculo-tendon system so it can generate the forces required to move the link-segment system. The *respiratory system* is another highly responsive system which adapts to current requirements on very short timescales. Its role is to provide oxygen and remove carbon dioxide, to enable the continuing metabolic processes powering the muscles. *Nutritional systems* operate over longer timescales, and their role is to provide the nutrients and energy required for the above systems. *Hormonal systems* operate over a wide variety of timescales, and are typically involved in cyclical regulation (e.g. circadian and seasonal rhythms, alternation between rapid-response-readiness and relaxation, etc.). Hormonal systems have broad effects across the other systems.

Internal (neuronal) resources

Where does the *nervous system* fit in this taxonomy? The nervous system responds to changes in its environment on millisecond timescales, which makes it suitable to serve as a fast-response system, and, we suggest, the primary medium supporting the informational (as opposed to direct, mechanical) couplings between the various subsystems. These need to be both stable and flexible, so that an organism can maintain informational couplings as long as needed, but de-couple and reform into a different device when the context changes. Structurally, then, the nervous system is *dynamically stable*, probably edge-of-chaos stable.[25] This stability does not reflect specific wiring connections being preserved, but rather the preservation of global function despite change in the underlying wiring, and it is actively maintained by the form of the informational flow. If you change that flow, the dynamic structure smoothly alters in response, as, for example, when you have someone pick up a tool,[26] or provide information suggesting a person actually has three hands.[27]

This structural flux still supports stable function; in biology, this capacity is referred to as *functional homeostasis* or, a bit oddly, as *degeneracy*.[28] This is defined as the capacity of a system to perform similar functions despite variation in how the functions are implemented.[29] The most remarkable demonstration of functional homeostasis in a nervous system comes from Prinz, Bucher, and Marder,[30] who simulated the functional output of the 20,500,000 different ways there are to combine the neurons involved in the pyloric (swimming) rhythm of the lobster's stomatogastric ganglion. Astonishingly, 20 percent of these combinations preserved pyloric-like activity (i.e. rhythmic swimming motions). Even when the simulations were constrained by data from actual lobsters, there were still 452,516 different ways to achieve the same functional outcome. Pyloric-like activity occurred for all 150 possible combinations of neurons, and for synapse weights across the entire functional range (with the exception of one particular synapse weight which had to be very weak: this matches the biology of the actual system very nicely). In effect, the behavior of the network exhibited extraordinary functional homeostasis (450,000 ways to produce the same rhythm) and the parameters were stable over several orders of magnitude (so long as the appropriate compensation occurred elsewhere in the network). Prinz et al. suggest this flexibility shows that the nervous system is *not* trying to maintain a specific set of wiring connections and weights; *the nervous system is trying to maintain the capacity to perform a specific function.* Sporns notes that "such homeostatic mechanisms are essential for the long-term stability of the brain given the continual remodeling and structural turnover of its cellular and molecular components."[31] Function is primary, and how that function is achieved is of secondary importance.

With some historic context and modern research insights in hand, we can now turn our attention to better explaining what something like the brain would do in this system.

A job description for the brain

Functional homeostasis seems to be an organizing principle that extends across multiple systems and levels in biology. The specific structural implementation is rarely that informative: It may simply be the solution that evolution and development happened to find first, by chance. *The real question of interest is: What is the function performed by the whole system, and how does the configuration of the system relate to the task?* This makes it all the more important that we adopt a task-specific approach[32] to understanding brain and behavior. A task-specific research strategy addresses the overwhelming and nonlinear "degrees of freedom" problem in the study of the human perception-action system. The proposed solution is to study the organization and composition of particular, carefully described examples of the human perception-action system at work, and then to describe the kinds of resources the system has access to for building such devices. In other words, figure out how a given task is achieved, and by doing so, come up with a list of things you know the system as a whole is capable of, which can then inform your study of related tasks. Typical model systems in perception-action research include coordinated rhythmic movement,[33] locomotion,[34] throwing and catching.[35]

A given perception-action task uses specific informational and motor components, and it explicitly does not use others. Formation of a stable perception-action system that will succeed in a particular task often requires that the task-relevant components become functionally "walled off" from irrelevant components (chewing gum doesn't interfere with walking, unless you are new at both). This enables the system to accomplish its goal stably and reliably for the duration of the task (i.e. by resisting perturbations).[36] Assembly is this process of coupling resources together to achieve some function. These resources are not mechanically connected, however (the way "the hip bone is connected to the thigh bone," for example), and they must therefore be coupled *informationally*.[37] This type of coupling affords the second critical aspect of assembly—*softness*, or the capacity to re-purpose your resources as required.

The nervous system is perfectly placed to implement these informational, temporary couplings between disparate resources. Information about these resources (both "internal" and "external") flows through the nervous system, which shapes itself in response. When this flow alters (i.e. when different task resources are available, or the same resources are in different states), the nervous system re-shapes itself in response, on millisecond timescales. This dynamically stable system is also, critically, everywhere in the body. So, in principle, if there is information about a resource flowing through the nervous system, it can be coupled to any other resource for which there is information.

This, then, changes the kinds of questions we ask about what the brain is doing. We are no longer interested in where and how the brain stores knowledge and learned skills, but in how the brain engages in the real-time coordination and control of behavior in response to stable but evolving patterns of information flow. To understand what the brain is doing, we need to first understand the forms of information to which it resonates. The ecological approach[38] is an example of a research program that is identifying one form of information (perceptual, typically visual information about the dynamics of events in the world). The answers to these questions should then inform the questions we ask about the brain. These new kinds of neuro-questions need a language with which to be asked, and that work remains to be done, hopefully with one eye on what pragmatism can contribute.

Connecting with pragmatism

The approaches discussed above have been developing in bits and spurts since the early 1900s as part of pragmatism's intellectual lineage within psychology. Let us translate some of the above to make this connection with pragmatism more obvious, acknowledging that we can do so in only a cursory fashion.

There is a lot of research within the ecological tradition looking at the catching of balls. Most people, most of the time, are not prepared to respond rapidly to incoming projectiles. We don't recommend you try this, but we assure you that if you randomly toss baseballs at strangers' heads, you will hit them in the head a lot of the time. In contrast, well-trained people, detecting a quick movement in their periphery and hearing a loud "Head's up!" will rapidly become a TSD specialized in catching things. Under such circumstances, people's accuracy in catching balls is remarkable given the precision required and the degrees of freedom involved. Imagine the ball is coming from slightly behind the person's side, right at the limits of their peripheral vision. Through a diffuse and highly flexible array of neural outputs, the orientation of the body shifts slightly. Included in this, the shoulder and back tense, pulling back the upper arm; the forearm naturally rotates upward without any central control needed, due to the configuration of bones and tendons. Similarly, the hand naturally opens. As the visual system engages more fully, fine motor control becomes coupled with the optical acceleration and optic expansion of the approaching objects so that the hand ends up at the correct location (with no "computation of the trajectory" required). When the flying object hits the center of the palm, the natural impact begins to close the fingers around it, and quick connections through the spinal cord finish the grasp. As you can see, the nervous system has a pervasive role in temporarily turning you into a ball-catching device, and it will similarly have a role in rapidly turning you into the next type of device as well, but it does so within the context of several nested systems.

But aren't there people who always seem ready to perform this task? Have we not heard of people making highly improbable catches of incoming projectiles before they even knew what they were doing? Yes, of course. Some people are prepared to perform certain types of tasks at more or less any moment. This is what Peirce called "habit," and his discussion of habit formed a central part of his influence on psychology. For Peirce, your habits, which also in some sense constitute your "character" and your "knowledge," are those things you do without hesitation. The person who is always at the ready to catch incoming objects is "that type of person" and clearly *knows* how to catch a ball; that vigilance is an important part of his or her *character*. Neuro*pragmatists* are thus obligated to answer the question: what is the role of the brain in the things you do without hesitation?

The modern research presented above, with some addition of Holt's vision toward explaining higher processes, lays a foundation for answering such questions, and more. Thus neuropragmatists are in a good position to help us find better ways of talking about (1) how we transform from one TSD to another; (2) how we develop the ability to become new types of devices; and (3) how we sometimes become rigid in our ability to switch between device-types. Or, to use Peirce's phrasing, we need ways to talk more accurately about the role of the nervous system in our developing and solidifying habits. This would be a bare beginning of connecting neuroscience and pragmatism.

An adequate language for this discussion will make explicit (1) the muti-leveled connections between the nervous system, the other bodily systems, and the wider world; (2) the circular nature of those connections, with some equivalent of a perception-action loop supporting behavior at each level; and (3) the tremendous redundancy in the nervous system and the wider system as a whole, which allows organisms to behave adaptively despite wide variations in the activity of the nervous system.

Conclusion

The nervous system develops, in both the short and long term, so as to preserve function, not structure. In order to understand the brain, or at least to talk about it correctly, we have to understand what those functions are, and we have to understand the many ways in which the nervous system can combine with other resources to fulfill those functions. The ecological approach proposes quite specific challenges that the organism has to solve: the organism must detect higher-order invariant patterns of ambient energy that specify action-relevant properties of the world, then use this action-scaled information to coordinate and control softly assembled TSDs. Behaviorist approaches to psychology, in contrast, tend to focus on the developmental support of behavior. While behaviorist approaches typically neglect physiology, they emphasize the dynamic nature of developmental process, and point out the incredible flexibility of the nervous system across environments. Neuroscience, in contrast, currently speaks a language derived from the cognitive approach, with its physiologized homunculi and rigid linear mechanisms. Reorienting cognitive science along more pragmatic, ecological lines alters the job description for the brain entirely, and redefines what we should be looking for in the brain. The above is a preliminary attempt to lay out a new way of conceptualizing what the brain is up to. It is time that the non-cognitive approaches to psychology stepped up and helped create a more honest way of engaging neuroscience.

Where to go from here? If you have read this far, the major impression we hope you have made is that there is a vocabulary vacuum that needs to be filled. Neuroscientists do not lack knowledge about the brain, and the psychologists do not lack knowledge about behavior. What we lack is a language to bridge the gap. We have so many things we want to say about what the brain does, and if we could articulate these things better, it would provide a more honest way to discuss the results of current neuroscience, and provide a framework to guide future neuroscience. However, we have not struggled with the problems long enough and deeply enough to have a solid, widely agreed upon vocabulary that can serve these purposes. Neuropragmatism, if pursued, will necessarily lead researchers to conclude that the conceptual apparatus of cognitive psychology is inadequate and misleading, especially when applied to the brain. Maybe, however, the pragmatists can stand and deliver where others have not. Maybe neuropragmatism can finally provide us with a sensible way of talking about the role of the brain in psychological processes.

Acknowledgments

This collaboration developed from ideas explored on the authors' blogs, "Fixing Psychology" (fixingpsychology.blogspot.com) and "Notes from Two Scientific Psychologists" (http://psychsciencenotes.blogspot.com). We wish to thank our readers, and especially those who have left comments and engaged us in discussion. If our page hits are any indication, there is a wide, international web of people interested in these ideas, and we are grateful that readers would find our contributions worthy of their valuable time and attention.

Notes

1 Lashley 1930, pp. 23–4, emphasis added.
2 MacCorquodale and Meehl 1948.
3 *Ibid.*, p. 106, emphasis added.
4 *Ibid.*, p. 105.
5 Charles and Dege 2008, p. 195.
6 Heft 2001; Smith 1986.
7 Wilson 2011.
8 E.g. Powers 1973.
9 Wilson and Golonka 2013.
10 Charles 2011a.
11 Holt 1936.
12 See also Charles 2011a.
13 For a more modern argument of similar points, see Tonneau 2011.
14 For a solid introduction to modern work on non-obvious causes of behavior, see Miller 1997.
15 Clark and Sokoloff 1999.
16 Raichle et al. 2001.
17 Bingham 1988.
18 Feldman 1966, 2011.
19 Bingham 1988.
20 Runeson 1977.
21 Gibson and Pick 2000.
22 Brooks 1997.
23 Bingham 1988.
24 Feldman 1966, 2011.
25 Alexander and Globus 1996.
26 Carlson et al. 2010.
27 Gutersham et al. 2011.
28 Mason 2010.
29 Sporns 2011.
30 Prinz et al. 2004.
31 Sporns 2011, p. 68.
32 Bingham 1988.
33 Bingham 2004b, 2004a; Haken et al. 1985.
34 Fajen and Warren 2003; Wilkie and Wann 2005.
35 Zhu and Bingham 2008; Fink et al. 2009.
36 Wilson and Bingham 2008.
37 Kugler 1986.
38 Gibson 1979; Golonka and Wilson 2012; Turvey et al. 1981.

References

Alexander, David M., and Globus, Gordon G. (1996), "Edge-of-Chaos Dynamics in Recursively Organized Neural Systems," in *Fractals of Brain, Fractals of Mind: In Search of a Symmetry Bond*, eds E. MacCormac and Maxim I. Stamenov. Amsterdam, Netherlands: John Benjamins Publishing Company, pp. 31–73.

Bingham, Geoffrey P. (1988), "Task-Specific Devices and the Perceptual Bottleneck," *Human Movement Science* 7(2–4): 225–64.

—(2004a), "Another Timing Variable Composed of State Variables: Phase Perception and Phase Driven Oscillators," in *Advances in Psychology, Vol 135: Time-to-Contact*, Heiko Hecht and Geert Savelsbergh (eds). Amsterdam, Netherlands: Elsevier Science Publishers, pp. 421–2.

—(2004b), "A Perceptually Driven Dynamical Model of Bimanual Rhythmic Movement (and Phase Perception)," *Ecological Psychology* 16(1): 421–2.

Brooks, Rodney A. (1997), "Intelligence without Representation," in *Mind Design 2: Philosophy, Psychology, Artificial Intelligence*, ed. John Haugeland. Cambridge, MA: MIT Press, pp. 395–420.

Carlson, Thomas A., Alvarez, George, Wu, Daw-an, and Verstraten, Frans A. J. (2010), "Rapid Assimilation of External Objects into the Body Schema," *Psychological Science* 21(7): 1000–5.

Charles, Eric P. (2011a), *A New Look at New Realism: The Psychology and Philosophy of E. B. Holt*. Piscataway, NJ: Transaction Publishers.

—(2011b), "Seeing Minds in Behavior: Descriptive mentalism," *Review of General Psychology* 15(3): 267–76.

Charles, Eric P., and Dege, Martin (2008), "The Beyond Must Also Be Useful: The burden of alternative approaches," *Integrative Psychological and Behavioral Science* 42(2): 194–9.

Clark, D. D., and Sokoloff, L. (1999), "Circulation and Energy Metabolism of the Brain," in *Basic Neurochemistry: Molecular, Cellular and Medical Aspects*, G. J. Siegel, B. W. Agranoff, R. W. Albers, S. K. Fisher, and M. D. Uhler (eds). Philadelphia, PA: Lippincott, pp. 637–70.

Fajen, Brett R., and Warren, William H. (2003), "Behavioral Dynamics of Steering, Obstable Avoidance, and Route Selection," *Journal of Experimental Psychology: Human Perception and Performance* 29(2): 343–62.

Feldman, Anatol G. (1966), "Functional Tuning of the Nervous System with Control of Movement or Maintenance of a Steady Posture, Ii: Controllable parameters of the muscles," *Biophysics* 11: 565–78.

—(2011), "Space and Time in the Context of Equilibrium-Point Theory," *Wiley Interdisciplinary Reviews: Cognitive Science* 2(3): 287–304.

Fink, Philip W., Foo, Patrick S., and Warren, William H. (2009), "Catching Fly Balls in Virtual Reality: A critical test of the outfielder problem," *Journal of Vision* 9(13): 14.

Gibson, Eleanor Jack, and Pick Anne D. (2000), *An Ecological Approach to Perceptual Learning and Development*. New York: Oxford University Press.

Gibson, James Jerome (1979), *The Ecological Approach to Visual Perception*. Boston, MA: Houghton, Mifflin and Company.

Golonka, Sabrina, and Wilson, Andrew D. (2012), "Gibson's Ecological Approach–a Model for the Benefits of a Theory Driven Psychology," *Avant* 3: 40–53.

Guterstam, Arvid, Petkova, Valeria I., and Ehrsson, H. Henrik (2011), "The Illusion of Owning a Third Arm," *PLoS ONE* 6(2): e17208.

Haken, H., Kelso, J. A., and Bunz, H. (1985), "A Theoretical Model of Phase Transitions in Human Hand Movements," *Biological Cybernetics* 51(5): 347–56.

Heft, Harry (2001), *Ecological Psychology in Context: James Gibson, Roger Barker, and the Legacy of William James's Radical Empiricism*. Mahwah, NJ: L. Erlbaum.

Holt, Edwin B. (1936), "Eight Steps in Neuro-Muscular Integration," in *Problems of Nervous Physiology and of Behavior*. Tiflis: Georgian Branch, Academy of Sciences, USSR, pp. 25–36.

Kugler, P. N. (1986), "A Morphological Perspective on the Origin and Evolution of Movement Patterns," *Motor Development in Children: Aspects of Coordination and Control*, pp. 459–525.

Lashley, K. S. (1930), "Basic Neural Mechanisms in Behavior," *Psychological Review* 37(1): 1–24.

MacCorquodale, Kenneth, and Meehl, Paul E. (1948), "On a Distinction between Hypothetical Constructs and Intervening Variables," *Psychological Review* 55(2): 95–107.

Mason, Paul H. (2010), "Degeneracy at Multiple Levels of Complexity," *Biological Theory* 5(3): 277–88.

Miller, David B. (1997) "The Effects of Nonobvious Forms of Experience on the Development of Instinctive Behavior," in *Evolving Explanations of Development: Ecological Approaches to Organism-Environment Systems*, Cathy Dent-Read and Patricia Zukow-Goldring (eds). Washington, DC: American Psychological Association, pp. 457–507.

Powers, William T. (1973), *Behavior: The Control of Perception*. Oxford: Aldine.

Prinz, A. A., Bucher, D., and Marder, E. (2004), "Similar Network Activity from Disparate Circuit Parameters," *Nat Neurosci* 7(12): 1345–52.

Raichle, M. E., MacLeod, A. M., Snyder, A. Z., Powers, W. J., Gusnard, D. A., and Shulman, G. L. (2001), "A Default Mode of Brain Function," *Procedings of the National Academy of Science, U.S.A.* 98(2): 676–82.

Runeson, Sverker (1977), "On the Possibility of 'Smart' Perceptual Mechanisms," *Scandinavian Journal of Psychology* 18(1): 172–9.

Smith, Laurence D. (1986), *Behaviorism and Logical Positivism: A Reassessment of the Alliance*. Stanford, CA: Stanford University Press.

Sporns, Olaf (2011), *Networks of the Brain*. Cambridge, MA: MIT Press.

Tonneau, Francois (2011), "Holt's Realism: New Reasons for Behavior Analysis," in *A New Look at New Realism: The Psychology and Philosophy of E. B. Holt*, ed. Eric Charles. Piscataway, NJ: Transaction Publishers, pp. 33–55.

Turvey, M. T., Shaw, R. E., Reed, E. S., and Mace, W. M. (1981), "Ecological Laws of Perceiving and Acting: In reply to Fodor and Pylyshyn," *Cognition* 9(3): 237–304.

Wilkie, Richard M., and Wann, John P. (2005), "The Role of Visual and Nonvisual Information in the Control of Locomotion," *Journal of Experimental Psychology: Human Perception and Performance* 31(5): 901–11.

Wilson, Andrew D. (2011), "What Does the Brain Do, Pt 2: The Fast Response System," in *Notes from Two Scientific Psychologists*. Available at: http://psychsciencenotes.blogspot.com/2011/08/what-does-brain-do-pt-2-fast-response.html

Wilson, Andrew D., and Bingham, Geoffrey P. (2008), "Identifying the Information for the Visual Perception of Relative Phase," *Perception and Psychophysics* 70(3): 465–76.

Wilson, Andrew D., and Golonka, Sabrina (2013), "Embodied Cognition Is Not What You Think It Is," *Frontiers in Psychology* 4(58): 1–13.

Zhu, Qin, and Bingham, Geoffrey P. (2008), "Is Hefting to Perceive the Affordance for Throwing a Smart Perceptual Mechanism?" *Journal of Experimental Psychology: Human Perception and Performance* 34(4): 924–43.

Part Three

Cognition, Inquiry, and Belief in the Brain and Beyond

8

How Inquiry and Method Shape Brain Science

Pragmatism, Embodiment, and Cognitive Neuroscience

Tim Rohrer

What view are we to adopt concerning impressions reproduced by mental causes alone, or without the aid of the original, as in ordinary recollection? ... The renewed feeling occupies the very same parts, and in the same manner, as the original feeling ... For where should a past feeling be embodied, if not in the same organs as the feeling when present? It is only in this way that its identity can be preserved; a feeling differently embodied would be a different feeling.

Alexander Bain, 1855[1]

It seems almost certain ... that the imagination-process differs from the sensation-process by its intensity rather than by its locality.

William James, 1890[2]

Pragmatism's sharp elbows

A member of the National Academy of Science once elbowed me in the ribs. As a young graduate student in philosophy attending a talk over in the psychology building, I had just asked another young graduate student presenter what might have been a devilishly hard question. I don't remember the question well, but what was whispered in my ear has always stuck with me: "That's the trouble with you philosophers. You always want to be right all the time. We're scientists over here—I know I'm wrong."

A more concise version of the pragmatic philosophy of science I have not encountered. Scientific truth is a matter of making distinctions that have useful, practical, significant, and long-term consequences, but it must also be a fallible process whose truths are subject to continual revision. Pragmatism, like my friend the neuropsychologist, has sharp elbows; that is to say that pragmatism places incompleteness, error, fallibility, perspective, and organismic purpose and change at the core of its theory of scientific knowledge and truth. Other philosophies of science typically construe such problems as problems for their theories of scientific truth, posing challenges for how a scientific description can possibly correspond to reality. Pragmatism, by contrast, sees these matters not as problems per se, but as constitutive of the process of truth-making. Scientific truths involve gaining footholds that allow us to see farther and usefully predict more of the future than before; they are not about gaining a perfect, Aristotelian perspective from which all truths can be seen and all things can be known.

How freeing it is to admit that one's scientific models can be—and indeed, will eventually be—proved wrong. It allows one the freedom to realize that scientific theorizing is essentially a speculative activity that is not identical with the object of study itself. Theory is not reality, errors in theories are to be expected, and perfection and completeness in description are beyond the ken of science. However, it is not the case that there are no better or worse theories, or models that fit the data better or worse. The art of science lies in being wrong in an interesting, productive, and useful way. That is what, on the pragmatist's view, differentiates a good scientific theory from one not as good.

I begin with this example of the sharp elbows as an embodied example of what it means to be doing science. As human beings with bodies, we all know what it feels like to be elbowed in the ribs, and as academics we all know what it is like to be on the receiving end of vibrant questioning of our intellectual work. (Hopefully, although perhaps not always, the latter is more pleasant.) But it makes for a useful and productive starting point, not only for any felt similarities between these two processes, but because understanding "embodiment" is precisely what cognitive science has failed to do in the past and at the forefront of what it is doing today. In this chapter, I trace this history of the new "neuropragmatism" through detailing some of the paradigmatic experiments, engineering successes and writings of the past century of North American research on cognition.

Changing minds: "Functionalism" from cheek by jowl to computer software

The danger lurking in the shadows of every discussion of the history of psychology and philosophy is that one will only see what one wants to see in the history and fail to understand how the same word might have come to mean very different things roughly a century later. The history of the term "functionalism" is a particularly rich example of this danger. Due to the influence of analytic philosophy of mind, in cognitive science the term has acquired precisely the opposite meaning as it originally had around the turn of the twentieth century. At the turn of the last century philosophical psychology—for the disciplines had not become particularly separate—was immensely concerned with methodological questions concerning the possibility of a "pure" philosophy or a "pure" psychology.

Within psychology, this debate was over whether psychology should consist purely of the analysis of the mental content of psychological phenomena, such as the analysis of dreams, the construction of self, or all of the "what" a person might be conscious of; or whether psychology should consist primarily of experimental analyses of the mental operations that make mental life, thought, or consciousness possible. If the latter was to be the road taken, the methodology would be to measure the bodily correlates of mental operations. At the outset of *Principles of Psychology*, William James argues that our mental life is essentially embodied, emphasizing that studies of brain and nervous system physiology is a necessary component of psychology: "Bodily experiences, therefore, and more particularly brain-experiences, must take a place amongst those conditions of the mental life of which Psychology need take account … a certain amount of brain-physiology must be presupposed or included in Psychology."[3] James's endeavor was to incorporate the insights of the associationist psychologist Alexander Bain into his method of measuring the bodily or "functional" correlates of the mind.

At the turn of the twentieth century the term "functionalism" in psychology meant the embodied method of study of mental operations. In his 1906 presidential address to the American Psychological Association, Dewey's former protégée James Rowland Angell summarized the theoretical commitments of functional psychology:

> We have to consider (1) functionalism conceived as the psychology of mental operations in contrast to the psychology of mental elements; or, expressed

> otherwise, the psychology of the how and why of consciousness as distinguished from the psychology of the what of consciousness. We have (2) the functionalism which deals with the problem of mind conceived as primarily engaged in mediating between the environment and the needs of the organism. This is the psychology of the fundamental utilities of consciousness; (3) and lastly we have functionalism described as psychophysical psychology, that is the psychology which constantly recognizes and insists upon the essential significance of the mind–body relationship for any just and comprehensive appreciation of mental life itself.[4]

To those who wanted a "pure" philosophical psychology that kept mental ideas neatly separate from the material world, the functionalist trend toward developing a science of mental operations via measuring the body's correlates of mental life was anathema.

In addition to the focus on mental operations and their bodily correlates, the functional approach to biology also viewed these mental operations as serving a biological purpose—namely the organism's own, and its evolutionary, survival. Angell acknowledges that this position was particularly inspired by philosophical pragmatism, which argued that our mental life—up to and including science itself—also serves in an evolutionary capacity as a successful adaptive strategy for humankind. Angell famously characterized the functionalist approach to psychology as one in which the psychologist and the general biologist were inextricably linked "cheek by jowl:"

> The functional psychologist then in his modern attire is interested not alone in the operations of mental process considered merely of and by and for itself, but also and more vigorously in mental activity as part of a larger stream of biological forces which are daily and hourly at work before our eyes and which are constitutive of the most important and most absorbing part of our world. … This is the point of view which instantly brings the psychologist cheek by jowl with the general biologist.[5]

By using the embodied cheek by jowl metaphor, Angell emphasizes that functional psychologists believed mental operations should be studied not just for their own sake, but also for the sake of understanding how these mental operations are inseparably attached to the organism's biological and evolutionary context.

However, functional psychology and its focus on mental operations was gradually supplanted in North America by the behaviorist school of psychology in the 1920s, and the early psychological functionalist's notion measuring the correlates of mental operations was almost entirely pushed aside in favor of a focus on measuring and predicting behavior, particularly in animals. Inquiry into mental life was left to philosophy, and eventually a philosophically inclined strain of behaviorism known as "logical behaviorism" emerged that would eventually morph into "functionalism" in analytic philosophy of mind. Logical behaviorism held that the content of mental states was really just possible behaviors and that mental states were knowable only by observing behavior; moreover, any talk of mental states was really just talk about possible behaviors.[6]

Thus the notion of "functionalism" in analytic philosophy of mind was focused on mental content rather than mental operations per se, as an empirical rival to the tradition of "pure" psychology found in fields such as psychoanalysis. As philosophical functionalism cast off the shadow of psychological behaviorism, its conception of mind became entirely disembodied and bereft of any direct connection to the bodily or evolutionary context in which a particular mind arose. While there are now far too many heads on the functionalist hydra to be precisely characterized by any single summary, philosophical functionalism argues that mental states are constituted by their causal relations both (a) to each other and (b) to the sensory input(s) which precede these mental states and the behavior(s) which stem from them (known as the "black-box" conception of the mind). Furthermore, philosophical functionalism seeks to explain how the content of mental states, such as the conscious experience of pain or of red, can be the same from one person to another (or, and more hypothetically, from human to alien, or human to computer). In most varieties of philosophical functionalism, such mental states can be instantiated (or "realized") not only by neural systems like human brains but can also be instantiated so as to have identical content in computers, Martians, on Twin Earth,[7] or any other number of substrates (known as the "multiple realizability," "substrate-neutrality" or "instantiation" hypothesis).[8]

Generally speaking then, philosophical functionalism holds that as mental states are identified by their functional role, they are able to be instantiated in various systems as long as the system performs the appropriate functions. This argument is most often given shape using a computer metaphor where the mind is software running on the brain's hardware. Just as computers are physical devices with an electronic substrate that performs computation on inputs to give outputs, so brains are physical devices, the mind is a neural substrate that performs computation on inputs which produce behaviors. Of course, functionalists who commit to multiple realizability also necessarily commit to a disembodied, acontextual theory of mind. For the mental content to be the same in a digital computer as in a human being, it would have to be content that is independent of any bodily constraints or evolutionary pressures. This disembodied conception of mind was the conception of functionalism that early cognitive scientists inherited; one precisely inverted from Rowland's view of psychology and biology linked together cheek by jowl.

Though in a few moments I will present evidence of mental operations that shows how untenable the lack of bodily constraints are in philosophical functionalism, I want to acknowledge first the appeal and utility of philosophical functionalism from a pragmatic perspective on the history and philosophy of science. With apologies to Hubert Dreyfus, the key insight is recognizing not so much "what computers can't do" but what computers can do much better than humans. Computers were designed to offload one particular kind of human cognition—performing arithmetic computations—onto a machine. For example, before computing machines had been invented the calculation of the ephemeris tables used in navigation was a time-consuming and laborious task performed by skilled human beings. One such person was Benjamin Peirce, the father of the early American pragmatist philosopher Charles Sanders Peirce; both were consulting mathematicians to the U.S. Naval Observatory and related agencies in the late nineteenth century. In the early twentieth century, the U.S. Naval Observatory was one of the agencies funding early research into computing machinery, recognizing the potential of offloading the labor of teams of human computers onto what are now known as punched card computers.[9]

In Europe, protecting ships during World War II provided a similar impetus to the development of computers. British military mathematicians were tasked with trying to break the German ciphers used by the German Navy and U-boats to sink Allied shipping supplying Britain. Recognizing that there were not enough qualified personnel to compute all the possible cipher combinations that could be set each day on the Enigma encoding machines, Alan Turing and other mathematicians set out to create a machine—the Bombe—that could compute ciphering combinations more rapidly than could an unassisted human being. Because the German cipher keys changed daily, the Bombes were designed to compute multiple cipher keys simultaneously, and by using multiple machines in parallel it was possible to decrease the amount of time it took before a human operator could determine that an intercepted message was intelligible and a cipher key had been found. Due to the constraints of life in wartime Britain, the British Tabulating Machine Company, was unable to manufacture enough Bombes. After joint cryptography efforts were approved in 1942, the Americans produced their own version of the Bombe at the United States Naval Computing Machine Laboratory in Dayton, Ohio. The director of that project, Joseph Desch of National Cash Register (NCR), was issued the first U.S. patent for an electronic calculator after the end of the war.[10]

Pragmatic philosophy of science holds that science, like other human endeavors, is motivated by our evolutionary and biological need to solve practical problems in order to survive and flourish. The calculation engines developed for peacetime navigation and wartime cryptography are clear examples of scientists solving practical problems involving survival. Similarly, the tabulating machines for the U.S. census and accounting applications that predated them are clear examples of developments designed to improve human flourishing. The importance of these successes should not be underestimated, for the success in modeling one very specific human behavior—arithmetic computation—is at the root of the appeal of philosophical functionalism. Without that success, and the funding that followed it, logical behaviorism and the functionalism of analytic philosophy it engendered would have likely remained an obscure footnote in academic thought. But because scientists and engineers had already had so much success in developing a computer to perform one of the most difficult human thinking tasks—and one that took mathematical acumen and training beyond the ken of many human beings—it seemed a small leap to modeling human cognition in general.

This success was inspirational in that it suggested that if computing machines could perform this important part of human cognition, all of human cognition might be performed by such machines. After the war ended, the British mathematician and philosopher Alan Turing posed the provocative question: "Can machines think?" He argued that, at least in theory, a computing machine could imitate a human being's command of language sufficiently well to be judged as human by an interrogator (the "Turing test"). Moreover, he argued that although "the nervous system is certainly not a discrete-state machine … [and] a discrete-state machine must be different from a continuous machine … the interrogator will not be able to take any advantage of this difference."[11] He concluded that "under these circumstances it would be very difficult for the interrogator to distinguish the differential analyzer from the computer."[12] The conditions of the Turing test were an impetus to the black-box and substrate-neutral approach to mind of philosophical functionalism, both in that only sensory inputs and behavioral outputs are present in the interrogation and in that the type of the machine (digital or continuous; or electrical, mechanical, or neural) does not matter.

Building on Turing's idealized conception of a universal computing machine,[13] the analytic philosopher of mind Hilary Putnam proposed a theory called machine-state functionalism in which any mind can be regarded as an instance of a Turing machine, and in which a mental state was logically equivalent to a Turing machine state. Differentiating the logical and physical realization of such machines, Putnam writes:

> In particular the "logical description" of a Turing machine does not include any specification of the physical nature of these "states"—or indeed of *the physical nature* of the whole machine. (Shall it consist of electronic relays, of cardboard, of human clerks sitting at desks, or what?) In other words, a given "Turing machine" is an abstract machine which may be physically realized in an almost infinite number of different ways.[14]

Notice that the success in modeling arithmetic computation—something a computing machine was designed to do better and faster than human minds do—had been first extended by Turing to modeling linguistic communication, and then by Putnam to modeling any mental state whatsoever. While Putnam's machine-state functionalism has largely been supplanted by other functionalist theories of mind, the claim that the machine can be physically realized in multiple ways—i.e. the multiple realizability hypothesis—is an important component of nearly all variations of philosophical functionalism. Even where it is explicitly or partly rejected, the multiple realizability hypothesis—and the underlying view that mental content is independent of its instantiation—still plays a significant theoretical role in philosophical functionalism. From the point of view of functional psychology and pragmatism's commitment to seeing human cognition in its biological and evolutionary context, it is most problematic that the mind has been robbed of any constraints from the physical body. In the next section I show why that is now an untenable position within cognitive science.

Imagination measured and the return of the absent body

Suppose we once again return from thought experiments concerning the independence of mental content to the matter of measuring mental operations. Are there any ways in which the body constrains our mental life? Consider a classic question about the nature of our imaginations: how do we imagine rotating an object in our minds, as opposed to by hand?

Most contemporary cognitive scientists and those in allied fields are familiar with a well-known psychological experiment on mental rotation by Roger

Shepard and Jacqueline Metzler.[15] Participants were asked to determine whether a two-dimensional drawing of a three-dimensional object was identical to or a mirror image of another such object. By measuring the time needed to answer the question, Shepard and Metzler found that participants mentally rotated the object at a linear rate—about 60 degrees per second. The more degrees a participant had to rotate the object, the longer it took to answer the question. In other words, participants appeared to be manipulating such images as wholes, preserving their topologies while mentally rotating them through a series of intermediate depictions. At the time of its publication, their finding was surprising because the prevailing computationalist and functionalist views of the time held that the mind operated in a symbolic rather than depictive fashion, arguing that any such mental imagery would be merely epiphenomenal.[16] Optimal search algorithms for a digital computer would operate quite differently by, for example, calculating the rotations at large degree intervals (such as 30, 60, 90 and so forth) and interpolating the results against the original to narrow down the search space in a coarse fashion before engaging in more finely tuned rotations.

That finding inspired a firestorm of controversy and variations on the initial experiment. One in particular is worth performing, both as a thought experiment and then again in reality. Imagine extending your right arm straight out in front of you, palm toward your face and pointing upwards. Now imagine rotating your hand, first 180 degrees to the left and then to the right. And then try to actually do it. Notice that the leftward (inward) rotation is relatively easy, while the rightward is quite difficult, requiring additional large shoulder and arm joint movements. A series of experiments by Parsons showed that when subjects were asked to perform mental rotations of images which consisted of line drawings of human hands instead of Shepard–Metzler 2D/3D block diagrams, subjects were quicker and better at identifying those rotations of the hand that were easier to perform given the bodily constraints on joint movements we all experience as humans. Furthermore, Parsons found that subjects were quicker and better at judging which hand—left or right—was pictured when imagining rotating the hand that did not require difficult bodily movements.[17] Given the details of how the body and its joints work, the motor imagery system actively constrains how fast the mental rotation is performed.

Consider also how patients with chronic arm pain in one arm perform similar mental hand rotation tasks. For their affected arm as compared to their uninjured arm patients are much slower to perform the necessary mental rotations in those conditions where the bodily movements that would be required for the actual hand rotation would require large arm movements. By contrast, a group of non-patient control participants showed no such differences between their left and right arms.[18] Not only does the body constrain how our mind works, but the body-in-pain constrains how the mind works.

Of course, a committed philosophical functionalist might argue that it is something specific to the mental content that determines whether the bodily constraints will apply. The argument would be something akin to the following: as these constraints emerge only when the experimental stimuli are not drawings of objects but are instead drawings depicting body parts, these constraints might only emerge from the set of relations between the mental content of multiple mental states (i.e. mental content concerning the motoric and joint relationships between hands, wrists and arms; and the arm pain for the second set of studies). Since the primary visual and motor systems are localized in functionally different regions of the cortex (and engage different sub-cortical loops), it would seem that differing stimuli might well naturally engage these differing regions of the brain cortex. Stimuli that depict body parts such as hands would engage the schema of the body from the motoric system that constrains the speed of mental rotation irregularly, while stimuli depicting small independent objects would be mentally rotated at a linear speed purely within the visual system. Since the mental content would presumably differ depending on which stimuli were presented to the participant's mind, a revised version of philosophic functionalism would still at least be plausible. This argument would then maintain that there is a "stimulus-driven choice" as to which neural subsystem does mental rotation.

Once again, however, the position of the philosophical functionalist is simply at odds with experimental results. Cognitive neuroscientists investigating whether the "choice" of which cortical subsystem is engaged as stimulus-driven have constructed three-dimensional objects from wooden blocks modeled on those depicted in the Shepard-Metzler diagrams. The actual 3D object was placed on a spindle defining the axis of rotation and just before the neuroimaging took place, participants either (1) watched the wooden blocks rotate on a motor-driven spindle—the visual system priming condition—or (2) turned the 3D objects by hand—the motoric system priming condition. Participants were then instructed to imagine rotating the visual stimuli presented during the neuroimaging in precisely the same manner. The neuroimaging results showed that the priming of each neural system resulted in increased activation in brain regions known to be associated with the respective neural subsystems. This result suggested that the immediately prior environmental experience, coupled with the social instructions, influenced whether a visual or motoric strategy was employed. In other words, the constraints are a result stemming not simply from the nature of the physiological body and limited to depictions of the physiological body but from the body interacting with the environment—both the physical environment in the physical interaction (i.e. viewing or manipulating) with the model stimulus, and the social environment in which the instruction to imagine it in the same manner was given.[19] The mental operations underlying mental rotation are subject to embodied constraints; this finding is precisely what the functionalists of analytic psychology denied and the functionalist psychologists in the tradition of early American pragmatism predicted.

Neural maps and a pragmatist theory of neural computation

A little more than a century after functional psychologists like James and Angell argued not only that mental operations were measurable but also that the mind and brain were organic biological entities that served the evolutionary needs of the organism (and its species), the pragmatist conception of mind as fundamentally embodied is once again becoming the normative working hypothesis of cognitive science. Although the computational models of early cognitive science sought to banish it, embodiment never really entirely departed the science of the "wet mind," or the more biologically based neuroscience of cognition—what we now term neuropragmatism. Although there are certainly some threads in the contemporary theory of neural computation that harken back to the disembodied problems of philosophical functionalism, in this section I show how even the conception of neurocomputation is fundamentally embodied in how it models the neural structures that underlie imaginative activities such as mental rotation.

One of the most profound findings in twentieth-century neuroscience is that neurons organize themselves into neural maps. While there are many sources of this discovery, neurosurgeon Wilder Penfield's experiments with patients under light anesthesia showed that the primary sensory and motor cortex contain fields of neurons that map the sensory or motor system of the body in a largely "topographic" fashion, for example with the wrist connected to the forearm and that to the elbow and the upper arm. As there are distortions and discontinuities in this map where there are large numbers of nerves and muscles (for example, the hand, face, and the genitals have dedicated cortical regions), Penfield proposed the size of the cortical region devoted to topographic organization was proportional to the importance of a bodily region to the organism's successful functioning.[20] Other brain systems have similarly direct mappings from experience to neuronal maps, such as the maps for pitch in the auditory cortex or the retinotopic maps of the visual cortex. Maps in brain regions that are active later in sensory processing are organized to use space to map increasingly abstract contours of perceptual experience and are more properly termed "topologic" than topographic. Operations like mental rotation take place in such neural maps, and the fact that there is a linear relationship between the number of degrees an object is mentally rotated and the amount of time it takes to perform the rotation in the original Shepard and Metzler experiment is a consequence of the cost of maintaining the perceptual contours of the mental image in neural maps in the visual system. But the fact that there are topographically organized neural maps is only part of the story; the other important part of this finding is that neural maps are self-organizing from interacting with stimuli during development.

While there are clearly genetic contributions concerning the cellular differentiation and migration patterns of neurons (e.g. Sperry argued that some neurons were genetically destined for retinotopic maps),[21] the chief principles behind the self-organization of groups of neurons into neural maps are epigenetic selective principles similar to the Darwinian principle of natural selection. Instead of selection by birth, there are several levels at which this selection operates competitively on neural populations. These proposals fall under the general rubrics of "neuronal selectionism"[22] or "neural Darwinism."[23] First, during organismic development there is a process of neuronal group selection during which groups of neurons migrate from the neural tube in the embryo to their final position in the organism's brain or spinal cord, some groups perishing along the way.[24] Second, there is axon-axon competition for space within a particular brain region as the axons grow into an array into order to form a continuous topographic target for sensory input, as "is seen not only in normal development, but even after fairly drastic experimental manipulations such as removal of half of the retina or tectum."[25] Finally there is a level of selection during which the synaptic boutons on the dendrites of a single neuron grow in response to the axonal firing of another, and wither away if there is no axonal input nearby.[26] These three mechanisms appear to work sequentially during the course of normal development, with the final stage of development usually considered to be a form of postsynaptic Hebbian tuning or refinement. (The degree to which Hebbian mechanisms should be considered "selective" has been controversial, but Narayanan and Johnston have recently shown that there can be functional maps that encoded in the spatial growth pattern of the dendrites, similar to the axon-axon competition for space between neurons.[27] Similarly, recent neurocomputational models suggest that a "selection" model may be a profitable way of looking at Hebbian learning.[28] By contrast, in the case of a traumatic injury, such as a cerebral infarct, there may be larger-scale neuronal reorganizations in primates that reinvoke multiple levels of selection (i.e. neural plasticity)[29] rather than relying on Hebbian tuning alone.)

One of the key series of experimental studies demonstrating how the perceptual contours of experience are selected for during the organism's development of such neural maps involves a line of experiments inspired by the North American stereotype of the wise old professorial owl, complete with glasses. Knudsen and collaborators[30] outfitted juvenile barn owls with sets of prismatic glasses that distort the visual field a number of degrees in one direction or another. The owls then grew to maturity having learned to hunt both wearing glasses and without them, using both the distorted information from the glasses and their natural vision. When the external nucleus of the inferior colliculus (ICX) of those owls was injected with an anatomical tracing dye, the researchers found an increased density of bifurcated axons with branches laden with synaptic boutons that extended only in the direction predicted given the direction of the prismatic distortion. In other words, the brains of the glasses-wearing owls showed two discrete neural maps wired into their neuronal arbors, one map for not wearing the glasses and the other map shifted topographically in the direction and in proportion to the angular distortion of their glasses.[31] Owls not outfitted with glasses when juveniles show no such bifurcation of axonal growth in the neural maps of their ICX. At least in the case of the barn owl's brain physiology, developmental learning becomes wired into the brain's neural maps by an experience-dependent process of neuronal selection.

The selective character of these "wet mind" studies of experiential learning is consonant with the evolutionary concerns of the functional psychology of the early pragmatists, but it is important for neuropragmatism to also recognize how the self-organization of neural maps can also be modeled computationally when the computational model is built in accordance with embodied constraints. Successes in the science of the "wet mind" approach have inspired many models more or less closely modeling the neurophysiological process, notably McCulloch and Pitts,[32] who built a computational model of neuronal axonal firing considered as a single computational unit. The degree to which neurocomputational models are verisimilitudinous to the underlying neurobiology varies, both as a function of the operational scale of the model and as a function of limits on raw computational power (since almost all neural networks are simulated on digital computers with discrete states).

However, there is always a continuum between "realistic" and "simplifying" models in any type of scientific modeling.[33] All models, by virtue of being models, are inevitably somewhat simplifying, and useful insights into mental operations can be had at different physical scales—so long as the modeling takes place at the appropriate scale. A neurocomputational model at the synaptic level of the ion channels and exchange of potassium and sodium can tell us something about the mechanisms through which interneuronal learning becomes hard-wired, but would be cumbersome for modeling the longitudinal development of experience-based learning like that which takes place over the entire life of the barn owl. Explanation requires a chain of such neurocomputational models, and models of physically larger or temporally longer phenomena will need to make simplifying assumptions about the way the underlying physiology works. Despite any simplifying assumptions such neurocomputational models are already anti-philosophical functionalist in that the model is at least partly determined by analogizing some of its computational processes to the underlying neurobiology, rather than presuming that the mental operations to be modeled are computationally independent of any such embodied constraints and can thus be instantiated or realized in any number of mediums.

The history of self-organizing neurocomputational models of the neural maps are a particularly instructive example of this productive interaction between the details of the neurobiology and the computational model. To take one of the most prominent threads in that research, Kohonen proposed that artificial neural network models of the topographic neural maps in the visual system could be improved by using a mathematical construct called a "neighborhood" activation function.[34] In the neighborhood function, the node whose weighting vectors have been increased in response to a stimulus also shares that increase with its nearest neighbor nodes, with the neighborhood activation trailing off as a function of distance. The result is that the nodes in the artificial neural network competitively develop a topographic spatial sensitivity similar to the observed neural maps, using a biologically plausible form of mathematics. While Kohonen explicitly states that his model is a mathematical model first and "is related to an idealized neural structure,"[35] his insight of a neighborhood function was based on earlier attempts to model the retinotopic maps found in the visual cortex.

Once again, however, the actual retinotopic neural maps that inspired Kohonen's neighborhood function were driven by a set of prior studies of mapping in the somatosensory cortex. In studies of the neural maps in monkey somatosensory cortices, Mountcastle had found that nearby neurons activate for sensory stimuli that are near (i.e. in the "neighborhood" of) the sensory stimulus that most excites it—e.g. a neuron optimized for a fingertip would be partially activated by stimulating the middle of that finger. In neurophysiological studies Mountcastle showed that the somatosensory cortex was organized in columns with a large pyramidal neuron extending vertically in the cortex and various types of interneurons wrapped about it. He proposed that the neighborhood activation effect correlated to the receptive field width of the various interneurons feeding the pyramidal neuron in that column. In his view, the functional organization of the cortical columns consisted of mini-columns organized into larger "hyper-columns," each composed of hundreds or thousands of neurons.[36] Mountcastle's research on the somatosensory system was then extended into the visual cortex by his colleagues Hubel and Wiesel, who found mini-columns retinotopically organized for line orientation selectivity and larger hyper-columns for ocular dominance.[37]

Mountcastle, Hubel and Wiesel's neurophysiological findings were subject to numerous neurocomputational modeling attempts, including the neural field model proposed by Amari for both retinotopic and somatotopic neural maps, who explicitly noted that "the sensors or effectors scattered on the body are mapped topographically (somatopy) [and] … such parts as fingers occupy a large area on the cortex so that they have good resolution."[38] Amari's attempt to develop a more naturalistic neurocomputational model of the neural maps observed in the neurophysiology of both modalities was the direct intellectual predecessor of Kohonen's better known work on idealized neurocomputational maps.

Over the ensuing years Kohonen networks have been criticized for simplifying the underlying neurophysiology excessively, particularly with respect to the lack of nodes modeling the interneurons of the cortical columns. For example, by adding a layer of nodes to model inhibitory firing by interneurons, an artificial neural network can attend to multiple visual features in a single map rather than the single visual feature Kohonen networks map.[39] Though Kohonen's idealized neighborhood function worked well enough, the underlying mathematics would undergo refinement to become more biologically plausible. Kohonen and Amari's practical achievements lay in their successful demonstrations of how the actual neural maps could form via self-organizing neurocomputational maps that take as inputs simple sensory stimuli. Even when the study of mental life is considered from a neurocomputational perspective, the functional psychology of the early American pragmatists triumphs over the functionalism of analytic philosophers—for the origin of the neurocomputation is not in Putnam-like Turing machines, but in the fleshy, embodied, wet science of neural maps.

Neuropragmatism as an embodied philosophy of cognitive science

In contemporary cognitive science and cognitive neuroscience, philosophical functionalism has been largely elbowed aside by two points that I have traced in this recounting of intellectual history. The two central insights of neuropragmatism should now be apparent: first, embodiment does matter to the measurement of mental operations as we have seen with the differences between the mental rotation of objects and of hands, the effects of the body-in-pain on mental rotation, and neuroimaging studies of where in the brain imaginative activity takes place. The twin epigraphs to this chapter from Bain and James uncannily anticipate the role of embodiment in contemporary cognitive science and cognitive neuroscience. Imagination and perception take place in the same brain areas, and the embodied effects of how we accomplish perception in our neural maps shape and constrain the action of our imagination.

Second, the brain is an organic system that grows and adapts by using selective principles that are (or are highly similar to) the evolutionary principles that the functional psychologists and the early pragmatists suggested. In developing self-organizing neural maps that adapt to environmental stimuli, the brain's embodiment does not resemble the type of discrete-state computer that was originally designed to solve difficult arithmetic and cryptographic problems. Modeling how regions of the brain self-organize into neural maps required designing a different type of computing machine, one whose mathematical structures were at least partially determined by closely emulating the neurophysiological embodiment of the brain. Moreover, the degree of oversimplification of such neurocomputational models due to the computational power necessary to model anything other than small regions of human cortex suggested that Turing's claim that any analog computer could be adequately simulated on a discrete-state one was both incredibly inefficient and unlikely to produce a machine that could pass the Turing test. The sheer scale of the problem has dented the optimism that gave rise to the functionalist dream of a thinking machine.

That dream has largely been replaced by something significantly more interesting: a vision of cognitive science that is biologically and evolutionarily informed by the underlying neurophysiology, that focuses on the embodied interactions of the organism and its environment, and, to the extent that the first generation's dream of a thinking machine survives, a growing realization that such a machine will need to be embodied as a robot with sensors and possibly even a morphology close to a human one. But the dream of the autonomous thinking machine has also changed in another respect—the thinking machine has literally become part of us, of our biology, with advances in creating artificial prostheses that can mesh with the nervous system, cochlear implants, and so forth. The fact that a patient can learn to exert conscious mental control over such innovations results from the successes in modeling how small-scale neural networks work in the brain and from successes in building prosthetics to which the remaining neurons can adapt.

As I write this chapter, I am keenly aware that the United States has been at war for over a decade, and the importance of that is not lost on me from a pragmatist perspective. Just as with the advances in surgical techniques that accompanied the Civil War, so have the advances in neuroscience, such as neurally grafted prosthetic limbs, recovery from traumatic brain injuries, knowledge of and treatment for the brain chemistry underlying post-traumatic stress syndrome, and so forth, resulted from the practical problems of treating patients suffering from the wounds of war and then reassimilating wounded soldiers back into civilian society. From the vantage point of the pragmatic history and philosophy of science, neither the political and scientific nor the social and technological are entirely separate fields of study; the purpose of even the most esoteric and "pure" neuroscience is ultimately still the whole patient's recovery of function; the human organism learning to flourish again in its new state with new tools. Our bodies are no longer simply thought of as one or more machines; instead, they are hybrids of organism and machine. Just as with the advances in computing precipitated by cryptographic needs in World War II, the rapid progress in neurocomputation is driven by medical necessity. Advances in neuropragmatism are in no small part due to the peculiar political, historical, and funding contingencies of the present era.

Conclusion

I am quite certain that I am not an adding machine. Nor am I a slide rule. So why should I be a "computer"? More to the point, why do we think the brain is a computer, and not an adding machine, an automobile, or another sort of machine? (Surely all of us as academics have sometimes come to the awkward realization that we are more like a television set at a party, droning on in the background, long after the attention of our audience has moved on.) My point is that a cognitive neuroscientist can be something of a mechanist and a materialist, including recognizing the value of computational and mathematical models, without committing to taking the brain-as-computer metaphor literally and buying the attendant information-processing model of the brain wholesale. The key lies in the neuropragmatists' realization that a computational (or neuro-computational) model is only a useful tool from the scientist's perspective. The model is simplified; it may be wrong; the map is never the territory.

Note that this confusion is somehow peculiar to theories of brain function. A meteorologist's computational model of the weather is not mistaken for the weather; the weather map is not mistaken for the weather itself; nor would we ever say that the weather itself is a "computer." It is a given that when meteorologists model the weather, the weather isn't performing computations even when it exhibits behavior that we are able to predict from mathematical and computational models. So why does the confusion arise so often in cognitive science?

The reason is that cognitive scientists all too often confuse their own models with the organism's own experience, particularly when they use language that fails to distinguish their standpoint qua scientists from the organisms qua organisms. Let me give an example. In an otherwise solid review article on neural maps, Hemmen makes this mistake: "In neural science, one of the presumably most often asked, and also by its content most fascinating, question reads: How does the brain represent the world surrounding it?"[40]

But why ask the question in this way? Why not ask: "How does the organism use its brain to navigate the world?" By framing the question from inside the scientist's standpoint ("In neural science …"), the problem becomes a matter of how the brain and its neural maps "represent" the world, and not of how the organism uses those maps. Putting the representational cart before the organismic horse (as Hemmen does) forces him into a puzzle by the end of his article, when he raises the question of "What are neuronal maps good for?"[41] Considered from the standpoint of functional psychology, the problem for the organism—Hemmen summarizes work on sand scorpions, paddlefish, and the barn owl—is one of seeing in order to move, hunt, and eat, not one of "representing" more or less accurately the outside world inside its brain. The organism simply wants to survive and flourish, and its internal neural maps are active, dynamic responses to stimuli that simply need to be good enough to keep eating. This is as equally true of human cognition as it is of non-human animal cognition.

If one still believes in a philosophy of science that takes a correspondence theory of truth as fundamental to scientific activity, then the talk of neural representation and "distorted" representations may seem innocuous. After all, scientists often subscribe to this view of truth and see themselves as trying to get their theories to correspond to the world, to accurately represent the data they gather in their experiments. But even if one holds that theory of truth, why should a scientist expect that the experimental subject is also trying to accurately represent the world in its cognition? That point applies whether the organism whose cognition is being studied is a human being or a sand scorpion. Accuracy and spatial "distortions" of the external world in the neuronal map are likely to be solutions to optimization problems stemming from the organism's needs, where the "distortions" that maximize edges and motion selectivity in the neural maps are but natural technology, innovations important to eating, to surviving, to flourishing. It is only from the scientist's perspective as an outsider observing the correspondences between the stimuli presented to the organism and the patterns of activation in its neural maps that there is a more or less accurate correspondence to the scientist's stimuli. The organism has no insight into the "quality" of its "representations" as seen from the scientist's perspective, other than it may perish if the scientist's interventions into its neural maps are so radical as to make it unable to hunt, eat, or move. Surviving and flourishing, not representation and correspondence, is king in the land of the brain when viewed from the neuropragmatist's perspective. Neural map-making is not governed by how well the "representations" in the map correspond to the world the scientists observe but by the instrumental value of the neural maps to the organism.

Neuropragmatism is thus much more than the realization that the neural embodiment of the particular brain structures being modeled matters to the mental content and mental operations they perform. It requires reconsidering the standpoint and perspective of the cognitive scientist in relation to their object of study, and recognizing not just that neurons self-organize but that organisms have their own goals and purposes that are not necessarily similar to those of the scientist. It requires a theory of truth that can accommodate embodied biological imperatives and implies a history and philosophy of science that is likewise provisional and open to revision. These requirements have their historical antecedents in American pragmatism and in psychological "functionalism" understood as the term was used at the turn of the twentieth century. Neuropragmatism is thus a rejoinder to the cognitive scientist's philosophical presuppositions about the scientific method and the nature of inquiry. In short, it is a sharp elbow in the cognitive scientist's ribs in order to make us remember what it is we do when we do science.

Notes

1 Bain 1855, p. 333. The quotation appears at length in James, 1890b, pp. 68–9. For discussion of Bain's writings anticipating the theory of neural networks, see Wilkes and Wade 1997, pp. 295–305.
2 James 1890b, vol. 2, p. 72.
3 James 1890a, vol. 1, pp. 4–5.
4 Angell 1907, pp. 85–6.
5 *Ibid.*, p. 69. The metaphoric expression "cheek by jowl" originates from Shakespeare's *A Midsummer Night's Dream*, Act 3, Scene 2.
6 For a summary from the perspective of analytic philosophy, see the account of twentieth-century philosophical psychology, which traces the transition of logical behaviorism into philosophical functionalism found in Mason et al. (2008), pp. 583–618.
7 Putnam 1973, pp. 699–711. Note that Putnam's Twin Earth thought experiment is often posed as an objection to functionalism, though it was originally intended to support Putnam's theory of semantic externalism. In it, the counterfactual posits that a planet called Twin Earth exists identical to Earth in every respect except that there is a water-like substance which is known by the same name and has all the same macro-level characteristics of water, but has a different chemical composition. Since all the relations between mental states, sensory inputs, and behavioral outputs concerning water (e.g. that water is wet), would be the same for a human being on Earth and their counterpart human doppelganger on Twin Earth, the content of the mental state of the human and their doppelganger would be the same whether instantiated here on Earth or on Twin Earth. However, the content of the Earth human's mental state of believing that water is wet differs from the doppelganger's because one belief is predicated on H_2O, while the doppelganger's belief is predicated on another chemical composition. The intra-theoretical objection here is that since the two people would be functionally identical, yet still have different mental states, functionalism cannot sufficiently account

for all mental states. Note what drives the debate is that the content of the mental state, the "what" of consciousness, must be (or not be) identical in both the human and their doppelganger. Putnam argues that this points to a definition of "meaning that ain't in the head" but instead is external to minds and brains.

8 Block 1996.

9 Pugh 2009. Similarly, another mathematically inclined person who tired of doing double duty as a "computer" was Herman Hollerith, who invented the first punch card tabulating machines for use in the U.S. Census. Hollerith went on to found the company that eventually became the Computing-Tabulating-Recording Co. (CTR), for whom the Peirce Accounting Machine Company was an early competitor. In 1921, CTR—by then renamed IBM—purchased the patents of the Peirce Accounting Machine Company. The Peirce Accounting Company was formed around the patents of John Royden Peirce, a Dayton, Ohio cousin of the Harvard Peirces who would become one of IBM's most prolific inventors—and was the reason why the U.S. Bombe project would take place in Dayton. See Cortada 2000.

10 Hodges 1983. See also Pugh 2009.

11 Turing 1950, p. 451.

12 *Ibid.*, p. 452.

13 Turing 1937, 230–65.

14 Putnam 1979, p. 371.

15 Shepard and Metzler 1971, pp. 701–3.

16 For instance, see Pylyshyn 1973, pp. 1–24.

17 See Parsons 1987a, pp. 178–241, 1987b, pp. 172–91, and 1994, pp. 709–30.

18 Schwoebel et al. 2001.

19 Kosslyn et al. 2001, pp. 2519–25; Kosslyn et al. 2002, pp. 635–42. See also Parsons et al. 1995, pp. 54–8 and Parsons 2003, pp. 515–51.

20 Penfield and Rasmussen 1950.

21 See discussion in Meyer 1998, pp. 957–80.

22 See Changeux et al. 1983; Changeux et al. 1973, pp. 2974–8.

23 Edelman 1987.

24 *Ibid.*, pp. 46–57.

25 Luo and Flanagan 2007, p. 288.

26 Edelman 1987, pp. 46–57. See also Luo and Flanagan 2007, p. 289.

27 Narayanan and Johnston 2012, pp. 2343–51.

28 See Fernando et al. 2008, p. 3775; Fernando et al. 2010, pp. 2809–57.

29 Edelman 1987.

30 Knudsen 1998, pp. 1531–3 and 2002, pp. 322–8. See also DeBello et al. 2001, pp. 3161–74.

31 DeBello et al. 2001.

32 McCulloch and Pitts 1943, pp. 115–33.

33 Sejnowski et al. 1988, pp. 1300–1, 1305.

34 See Kohonen 1982, pp. 59–69 and 1984.

35 Kohonen 1982, p. 59. This is in direct contrast to earlier neurocomputational models based more rigorously on the actual neurophysiology, such as that done by Shun-Ichi Amari (1980), in "Topographic Organization of Nerve Fields," *Bulletin of Mathematical Biology* 42: 339–64.

36 Mountcastle 1957, pp. 408–34.

37 Hubel and Wiesel 1959, pp. 574–91.

38 Amari 1980, p. 339.

39 See discussion in O'Reilly et al. 2000, pp. 105–6, 236.

40 Leo 2002, p. 291.

41 *Ibid.*, p. 297.

References

Amari, S.-I. (1980), "Topographic Organization of Nerve Fields," *Bulletin of Mathematical Biology* 42: 339–64.

Angell, J. R. (1907), "The Province of Functional Psychology," *Psychological Review* 14: 61–91.

Bain, A. (1855), *The Senses and the Intellect*. London: John W. Parker and Son.

Block, N. (1996), "What is Functionalism?" in *The Encyclopedia of Philosophy Supplement*, ed. Donald M. Borchert. New York: Macmillan.

Changeux, J-P., Courrège, P., and Danchin, A. (1973), "A Theory of the Epigenesis of Neural Networks by Selective Stabilization of Synapses," *Proceedings of the National Academy of Sciences USA*, 70.

—(1983), *L'Homme Neuronal*. Paris: Fayard (trans. *Neuronal Man: The Biology of Mind*, New York: Pantheon, 1985).

Cortada, J. W. (2000), *Before the Computer: IBM, NCR, Burroughs and Remington Rand and the Industry They Created*. Princeton, NJ: Princeton University Press.

DeBello, W. M., Feldman, D. E., and Knudsen, E. I. (2001), "Adaptive Axonal Remodeling in the Midbrain Auditory Space Map," *Journal of Neuroscience* 21: 3161–74.

Edelman, G. (1987), *Neural Darwinism: The Theory of Neuronal Group Selection*. New York: Basic Books.

Fernando, C., Goldstein, R., and Szathmáry, E. (2010), "The Neuronal Replicator Hypothesis," *Neural Computation* 22: 2809–57.

Fernando, C., Karishma, K. K., and Szathmáry, E. (2008), "Copying and Evolution of Neuronal Topology," *PLoS ONE* 3.

Hodges, A. (1983), *Alan Turing: The Enigma*. New York: Walker and Co.

Hubel, D., and Wiesel, T. (1959), "Receptive Fields of Single Neurons in the Cat's Striate Cortex," *Journal of Physiology* 148: 574–91.

James, W. (1890a), *Principles of Psychology*, vol. 1. New York: Henry Holt.

—(1890b), *Principles of Psychology*, vol. 2. New York: Henry Holt.

Knudsen, E. I. (1998), "Capacity for Plasticity in the Adult Owl Auditory System Expanded by Juvenile Experience," *Science* 279.

—(2002), "Instructed Learning in the Auditory Localization Pathway of the Barn Owl," *Nature* 417.

Kohonen, T. (1982), "Self-Organized Formation of Topologically Correct Feature Maps," *Biological Cybernetics* 43: 59–69.

—(1984), *Self-Organization and Associative Memory*. New York: Springer.

Kosslyn, S. M., Ganis, G., and Thompson, W. L. (2002), "Neural Foundations of Imagery," *Nature Reviews Neuroscience* 2.

Kosslyn, S. M., Thompson, W. L., Wraga, M. J., and Alpert, N. M. (2001), "Imagining Rotation by Endogenous Versus Exogenous Forces: Distinct neural mechanisms," *NeuroReport* 12: 2519–25.

Leo, H. J. (2002), "The Map in Your Head: How does the brain represent the outside world?" *ChemPhysChem* 3: 291–8.

Luo, L., and Flanagan, G. (2007), "Development of Continuous and Discrete Neural Maps," *Neuron* 56: 286–300.

Mason, K., Sripada, C., and Stich, S. (2008), "The Philosophy of Psychology," *The Routledge Companion to Twentieth-Century Philosophy*, ed. D. Moral. London: Routledge, pp. 583–618.

McCulloch, W. S., and Pitts, W. (1943), "A Logical Calculus of the Ideas Immanent in Nervous Activity," *Bulletin of Mathematical Biophysics* 5: 115–33.

Meyer, R. L. (1998), "Roger Sperry and His Chemoaffinity Hypothesis," *Neuropsychologia* 36: 957–80.

Mountcastle, V. B. (1957), "Modality and Topographic Properties of Single Neurons of Cat's Somatic Sensory Cortex," *Journal of Neurophysiology* 20: 408–34.

Narayanan, R., and Johnston, D. (2012), "Functional Maps Within a Single Neuron," *Journal of Neurophysiology* 108: 2343–51.

O'Reilly, R. C., Munakata, Y., and McClelland, J. L. (2000), *Computational Explorations in Cognitive Neuroscience*. Cambridge, MA: MIT Press.

Parsons, L. M. (1987a), "Imagined Spatial Transformations of One's Hands and Feet," *Cognitive Psychology* 19: 178–241.

—(1987b), "Imagined Transformations of One's Body," *Journal of Experimental Psychology* 116: 172–91.

—(1994), "Temporal and Kinematic Properties of Motor Behavior Reflected in Mentally Simulated Action," *Journal of Experimental Psychology: Human Perception and Performance* 20: 709–30.

—(2003), "Superior Parietal Cortices and Varieties of Mental Rotation," *Trends in Cognitive Science* 7: 515–51.

Parsons, L. M., Fox, P. T., Downs, J. H., Glass, T., Hirsch, T. B., Martin, C. C., Jerabek, P. A., and Lancaster, J. L. (1995), "Use of Implicit Motor Imagery for Visual Shape Discrimination as Revealed by PET," *Nature* 375.

Penfield, W. G., and Rasmussen, T. B. (1950), *The Cerebral Cortex of Man*. New York: Macmillan.

Pugh, E. W. (2009), *Building IBM: Shaping an Industry and Its Technology*. Cambridge, MA: MIT Press.

Putnam, H. (1973), "Meaning and Reference," *Journal of Philosophy* 70: 699–711.

—(1979), "Minds and Machines," in *Mind, Language and Reality: Collected Papers*, vol. 2. New York: Cambridge University Press.

Pylyshyn, Z. W. (1973), "What the Mind's Eye Tells the Mind's Brain: A critique of mental imagery," *Psychological Bulletin* 80: 1–24.

Schwoebel, J., Friedman, R., Duda, N., and Coslett, H. B. (2001), "Pain and the Body Schema Evidence for Peripheral Effects on Mental Representations of Movement," *Brain* 124: 2098–2104.

Sejnowski, T. J., Koch, C., and Churchland, P. S. (1988), "Computational Neuroscience," *Science* 24: 1299–306.

Shepard, R. N., and Metzler, J. (1971), "Mental Rotation of Three-Dimensional Objects," *Science* 171.

Turing, A. M. (1937), "On Computable Numbers, with an Application to the Entscheidungsproblem," *Proceedings of London Mathematical Society, Series 2*, 42: 230–65.

—(1950), "Computing Machinery and Intelligence," *Mind* 49: 433–60.

Wilkes, A., and Wade, N. (1997), "Bain on Neural Networks," *Brain and Cognition* 33: 295–305.

9

Extended Mind and Representation

F. Thomas Burke

Good old-fashioned cognitive science characterizes human thinking as symbol manipulation qua computation and therefore emphasizes the processing of symbolic representations as a necessary if not sufficient condition for "general intelligent action."[1] Recent alternative conceptions of human thinking tend to deemphasize if not altogether eschew the notion of representation.[2] The present chapter shows how classical American pragmatist conceptions of human thinking can successfully avoid either of these extremes, replacing old-fashioned conceptions of representation with one that characterizes both representatum and representans in externalist terms.

First, cognitive externalism will be briefly explained, along with a review of how a pragmatist conception of experience stacks up as a kind of "pre-cognitive" externalism.[3] Second, a pragmatist social-psychological conception of mind will be presented—along lines initially developed by George Herbert Mead and John Dewey—to show how both symbol systems and systems symbolized (mind and world, ideas and facts) may be conceived in externalist terms.[4] Finally, an externalist conception of representation along these lines will be briefly outlined. Each of these three steps supports an alternative version of the brain-as-computer metaphor where the brain and nervous system are cast as only *part of* a computer, e.g. as a so-called *chipset*.[5]

Pragmatist psychology and cognitive externalism

Cognitive externalism is more an attitude than a particular doctrine. As such, it is a background framework of assumptions or presuppositions pertaining to how best to go about investigating the nature of human cognition.

Cognitive externalism emerged at the end of the twentieth century as part of a response to the research methodology of "good old-fashioned artificial intelligence and robotics" (GOFAIR)—a multidisciplinary research paradigm that prevailed from the 1950s into the late 1980s. The GOFAIR conception of mind was taken essentially from seventeenth- and eighteenth-century British empiricism by way of the psychophysical methodology pursued in the "new psychology" of the latter half of the nineteenth century. In the 1950s, *computers* and *computation* and a respective computer-metaphorical conception of the brain and/or mind emerged—basically as a supercharged version of the chemistry metaphor at the heart of Lockean empiricism and the steam engine metaphor at the heart of Humean empiricism, not to mention the clock metaphor at the heart of Cartesian rationalism. Unfortunately, GOFAIR research did not purposefully confront the recalcitrant conundrums of modern epistemology, hoping perhaps that a computer metaphor would be able to bypass if not cut through those conundrums. The GOFAIR paradigm therefore languished by the end of the 1980s, having failed to measure up to its own high expectations. A number of alternative perspectives began to take hold promoting various conceptions of mind as embodied, embedded, enactive, and extended,[6] not to mention situated, distributed, dynamical, and ecological.[7] Computation and computer metaphors have remained in the picture, but recent innovations have been less constrained by the suppositions and presuppositions of GOFAIR.

In particular, Clark and Chalmers published their seminal article, "The Extended Mind,"[8] that initiated a debate about the plausibility of an externalist conception of mind. Cognitive externalism is essentially a kind of functionalism, drawing on the principle of multiple realizability[9] as reflected in what Clark and Chalmers call the "parity principle." In short, any part of the world that functions in a process that we would easily accept as part of a cognitive process if it were to take place in the brain may be just as easily accepted as being part of a cognitive process.

In effect, the extended-mind thesis may be cast as a version of the computer metaphor except that the "motherboard" is allowed to encompass aspects of the world beyond the brain and central nervous system. For example, the brain might function on the whole or for the most part as a "chipset" with a nervous system composed of various "buses," while the mind's "CPU" and "memory" may include aspects of the body *and* its environment. What matters is the functionality of the system as a whole regardless of how and where its structural realization is located. If information that is present in the world can be accessed and utilized as easily and "immediately" as is information present in neural systems, then there is no good reason to characterize only the latter as accessing "main memory."[10]

The implications of such a view are significant in various ways. The point to be emphasized here is that a similar view was promoted by Mead and Dewey more than a hundred years ago in a different context, preceding by many decades the advent of the computer metaphor and machine-state functionalism. The Chicago school of pragmatism in the 1890s promoted a form of functionalism, after all, though functionality was conceived in terms of the utility of various actions in instances of agents having to deal with given uncertainties or present difficulties. This conceptual orientation was Darwinian, not computational. The agent in this case was at bottom a live creature instinctively intent on staying alive. Episodes of resolving difficulties by a given live creature are what Dewey called "experiences"[11] and what Mead called "acts."[12] These notions are directly related to Dewey's early notion of the "unit of behavior"[13] and his later conception of inquiry,[14] which in turn is an enhanced generalization of Peirce's doubt-belief conception of inquiry[15] as a process of "fixing" (as in stabilizing) belief in response to some respective destabilizing doubt (where scientific methodologies supply the best way to find such solutions etc.). The ongoing procedural nature of staying alive conceived in such terms was therefore neither computational nor non-computational in character, nor did it assume mental or representational capabilities in its most general formulation.

To put it simply, the structuralist distinction between sensory input and motor reaction (together constituting a "reflex arc") was to be replaced by a functionalist stimulus/response distinction. The simplest "unit of behavior" (Dewey) or the simplest "act" (Mead) was to be identified with a process of habituation or conditioning (a process of resolution etc.) in which (1) interactive but discordant sensory inputs and correlative motor outputs are *both* present in the stimulus, and where (2) the response is a learned (educated, evolved) coordination of such input/output interactions—the (re)formation of a particular routine mode of behavior (a habit, as it were) in which respective inputs and outputs are reliably if not robustly coordinated. *An experience* (for Dewey) or *an act* (for Mead) is thus neither a sensory input nor a motor output but rather a distinct episode of resolution (habit formation, learning, etc.) in which discordant manners of input/output interaction are ultimately and routinely coordinated. (Or when such coordination is not forthcoming, exit from and avoidance of such interactive circumstances is perhaps the only "resolution," if such remains possible.)

In this view, things are what they are for a given live creature by virtue of their functional roles in resolving difficulties. A brick or a shoe *is* a hammer not by virtue of its particular matter or general form but rather because it can fulfill the function (can serve the purpose) of pounding a nail. Not just mental abilities but *any type of thing* would be cast in such functional terms. Objects are what they are (on the fly) by virtue of how they function in service to living—to one's staying alive, maintaining some degree of well-being if not perhaps flourishing.

The fact that this view is compatible with a kind of cognitive externalism is not hard to see. First, consider the claim that an experiencer (and thus a thinker?) is *constituted* by systems of habits—stable systems of processes and mechanisms for orchestrating input/output interactions. If such habits are extended, then an experiencer constituted by such habits is extended. So are habits extended? Habits consist of reliable ways that systematic couplings of both inner and outer events and processes are coordinated and wielded in respective inner/outer circumstances. As embodiments of law-like regularities, habits are relational in character. Peirce[16] describes habits as general law-like "rules of action." As such, habits are engrained attunements of the experiencer to relationships among variables such that changes in one or more such variables will, in a regular law-like manner, involve (effect or be the effect of) variations in other respective variables. These law-like regularities may involve both "inner" and "outer" variables, not to mention variables that are not exclusively either.

For instance, a child's learning how to walk depends as much on reliable constancies in its environment (gravitation, flat surfaces, hard surfaces, etc.) as on reliable physiological features (neural, muscular, skeletal, etc.). The presence of a constant force of gravity is as essential to learning to walk as is the constant rigidity of one's shins and thighbones. A direct law-like correlation between energy expended and distance travelled can be realized only in the presence of such inner and outer constancies. Thus a learned ability to walk cannot be located only in growing muscles and bones (involving e.g. bone rigidity, energy consumption, the coordination of visual and tactile neural systems, etc.) but must involve respective variables in the environing world as well (e.g. the direction and magnitude of the force of gravity, directions and distances to be travelled, the orientation and contour of the surrounding terrain, etc.). Some parts of the walking process will also involve variables that singularly cannot be so easily identified as bodily or environmental (e.g. directions and speeds of limb movements, directions of bodily orientation relative to the direction of the force of gravity, etc.). This is one of any number of such examples of habits that incorporate both bodily and environmental variables. We may readily say that habits as such involve (causal, dynamic) *couplings* of processes and mechanisms involving both environmental and bodily variables.

If the experiencer is constituted by habits in this sense, then the experiencer would be extended across any alleged inner/outer divide as far as its habit-couplings are extended across any such divide. An experiencer will be *constituted* by inner/outer couplings *insofar as* it is constituted by its habits.

But there is a glitch here. The coupling/constitution challenge[17] is not yet answered merely by saying that the experiencer *is* the couplings. That conclusion has been drawn by way of arguing that habits are extended; but there is no principled reason (yet) not to say instead that the experiencer *is* only those aspects of habits that are enclosed within the skin etc., not including anything else in the external world that may be necessary for their proper functioning.

In response, we may instead postulate that an experiencer is constituted not by habits as such but by the *continued functioning* of (systems of) habits. That is, the one most fundamental fact about an experiencer is that it is alive and thus continually active in various ways, living in ways governed by its habits and frequently if not forever attending to instances where the ongoing functioning of those habits is impeded or threatened. These life-activities *constitute* the experiencer. That is, an experiencer is made up not of habits, strictly speaking, but of the *workings* of habits.

As such, an experiencer is an integrated (and growing) fusion of experiences—that is, experiences regarded in a count sense as episodes of coordination etc. The word "experience" is also used as a mass noun ("His experience was limited to advertising and sales") or as a verb ("She experienced success for the first time"). In all of these kinds of usage—whether as a count or mass noun, or as a verb—the core notion is that of coordination of an *extended* range of causally coupled things and activities.

If this sounds odd or unusual, consider the transitive verb "plowing" ("He is plowing the north forty"), the intransitive verb "plowing" ("He is plowing quickly"), the mass gerund "plowing" ("Plowing takes time") and even perhaps a count gerund "plowing" ("His plowings from one day to the next were consistently straight and even"). Implicated in each of these kinds of usage is an activity involving e.g. a tractor pulling a plow across a field with the curved plow blades slicing through and turning over portions of soil, all dependent upon a uniform downward gravitational force making possible the traction of the tractor tires as well as the slicing force of the plow blades. One might ask then where the plowing as such is located in this scenario. In one legitimate sense, it is where the plow is actually slicing the soil. That might be the focus of a mechanical engineer intent on designing a better plow. Note that plowing in that sense is not just in the soil nor just in or on the blade but in the *slicing* of the soil by the blade. Likewise, for a farmer intent on *using* a given plow to prepare for this year's crop, plowing may be regarded as the regular back-and-forth movement of the tractor and plow through the field—located not in any one component of this activity but in the activity as a whole (including the continuing pull of gravity keeping the tractor and plow pressed to the soil). In a third sense, plowing can be regarded as a "simple" act as part of growing and selling a crop—regarded, namely, as a black box where the "box" encloses an extended range of activities as outlined in the previous two senses of the word. The word "plowing" has many such senses, but it is easy to see that, whether as a noun or a verb, it designates a kind of working that is extended across various causally coupled things and activities.

Analogously, the workings of habits that constitute an experiencer are easily conceived, in perhaps several senses, as extended across many causally coupled things and activities inside and outside of the head.

To summarize, this position may be characterized as follows:

1. the experiencer is first and foremost a *live, sentient creature* and thus a fusion of token life-activities;
2. these life-activities are systematically governed in law-like ways by *habits* (as explained above);
3. the creature's *efforts* and *attention* tend to be directed at conditions where such habits are functioning with some actual or prospective difficulty;
4. the creature's *experiences* are (by definition of the term as a count noun) episodes of attempting to resolve such difficulties;
5. generically, the *purpose* of any such experience will be to resolve the respective discordance or difficulty so as to maintain some overall coordination of the experiencer's habits (even if some of the latter require modification); and
6. distinctive elements or features or dimensions of an experience are what they are (contextually) by virtue of the type of role they play in securing a resolution to the given difficulty.

The last claim says in effect that this position is a kind of externalist-friendly *functionalism*. On this account, an experiencer is continually repairing itself (or attempting to do so) from one experience to the next. The point of such repair is to secure and maintain the coordinated working of inner/outer life-activities insofar as that coordinated working is the experiencer itself. In that case, different aspects of experience are what they are by virtue of the role they play in such working. In *this* functionalist sense, moreover, as the coordinated working of inner and outer life-activities, the experiencer is best regarded as extended. That is, the experiencer *is* the working of habits, and *the working of habits* is extended.

This includes the workings of life-activities before, during, and after any given episode of repair. It includes the workings of repair itself, the latter being the essence of *experience* (in a mass sense), episodes of such repair being the *experiences* (in the count sense) where attention is focused (where effort is directed etc.). The coupling/constitution issue in the case of experiences (as opposed to the experiencer) is a matter of locating where such repair processes take place. Item (5) in the summary above indicates how the actual repair process will often be extended into the world (building a fire, patching a roof, locating water). What matters in such instances is securing the working coordination of inner and outer processes such that repairs will typically include manipulations of environmental factors involved in that coordination.

Applying a computer metaphor here, we should expect that the workings of the "CPU" of an actual (as opposed to artificial) experiencer will be slow compared to those of present-day computers given that the former will sometimes involve, e.g. tactile/mechanical manipulations of objects in the environment (pebbles in grooves of wooden or stone tablets, chalk marks on blackboards, pencil or ink marks on pieces of paper, and so forth). Actual calculators and computers in general (not to mention other technologies), rather than *modeling* brains, in fact provide mechanical *enhancements* of the latter kinds of manipulations (speed them up, render them more reliable etc.) and thus, as external devices, may become incorporated into the constitution of experiencers who can effectively utilize them.

In this sense, rather than think (metaphorically) of brains as computers, it might be better to think (literally) of computers as machines that enhance our natural abilities to manipulate information. The brain is no more a computer than it is a clock or steam engine. Steam engines, with some added engineering, enhance what we can do with our arms, hands, backs, and legs (as conduits of power). Computers enhance what we can do with pebbles in grooves or pencil marks on paper (as conduits of information). A computer metaphor may well be useful, but things need to be aligned and characterized properly. A brain *by itself* is not best thought of as a CPU. Rather, the workings of an actual CPU, literally, are a technological enhancement of *manual activities* that otherwise are slow, monotonous, and respectively unreliable. The brain, from a metaphorical perspective, is more like a "chipset," mediating processes that take place outside of the brain/chipset itself—processes of a manual character that take place outside of the body altogether.

So, then, what about *mental* abilities if we do not liken them to CPU-centric computational abilities? The word "mind" is not a gerund. The word "thinking" (as a verb or as a gerund) is closer to "plowing" in grammatical usage. One could argue by analogy that thinking is extended in various senses much like plowing is extended (as two different but analogous kinds of workings) if only one could specify the analogs of tractor, plow, soil, and/or gravitational force. Actually, whether the latter is doable or not, it misses an interesting point.[18]

Grammatical usage of the word "mind," as a noun, is more like the grammatical usage of the noun "home." Many uses of the noun "home" also implicate various kinds of "workings" that are hard to locate anywhere that is not extended across various causally coupled things and activities. When asked where your home is, you may first think of a particular house, though as the song says, "a house is not a home when," etc. When a real estate company tries to sell you a "home," they are really only selling you a house, though the sales pitch implicates that the house is such that it could be the center or focus of activities that constitute a home. A house may well be the locus of a home, but it does not encompass the full extent of a home insofar as the wider neighborhood and community (including everything from schools and roads and accessible employment opportunities to reliable plumbing and power) is important if not essential to maintaining the kinds of "holdings tight" and "kisses goodnight" that also are essential to "turning a house into a home." Might we regard *mind* in an analogous way?

The interesting point here is that we have already cast experience and experiencers in just this way. It is not clear what more can be said along these lines that would distinguish *mind* over and against *experience*. It might be that there is no real difference other than that mind (something possibly unique to humans) is just a particularly complex mode or faculty of experience that has emerged over millions of years of natural evolution. Thus mind is extended because experience is extended. But this is not very satisfying even if there is perhaps a grain of truth in it. Several distinctive features of thinking remain unaccounted for if we stop with merely a hand-waving appeal to evolutionary "complexity."

These missing features are linked to the notion of representation. They are features that are perhaps best explained in terms of representation and the use of representations. For instance, the word "think" is a verb ("You should think about what you have done"). Thinking is a kind of activity that has a distinctive "aboutness." We think *about* stuff. This is something else besides the directedness of experiences. That is, we think about stuff typically for some reason. Having a purpose, a reason for doing such and so, is a way of being directed. Thinking might take place in order to achieve that purpose, e.g. as a way of stepping back and planning a course of action or assessing possibilities. Another way to illustrate this point—that aboutness and directedness are not the same thing—is to contrast thinking with inquiry. Inquiry typically involves more than just thinking, but thinking may take place in order to advance some inquiry. Inquiry will be *directed* at some problematic subject-matter, while respective thinking will be *about* features of that subject-matter. For instance, one may think about the facts at hand in order to formulate some testable explanation of those facts.

Nothing in the last paragraph is particularly contentious, though it does not tell us very much about what thinking is. Whether we regard mind as extended or not, we need to be able to make sense out of the fact that we sometimes stop what we are doing and think about it. We sometimes stop and think about what we are doing. We sometimes stop to think. There are various ways to say that. It casts at least some kinds of thinking as involving disengagement from ongoing activities in order to reassess aspects of those activities that "give one pause," as it were. Casting experiencers as bundles of workings of intertwined habits does not directly say anything in particular one way or the other about this kind of disengagement that accompanies at least some kinds of thinking.

A plausible explanation might be found in an account of representation and the use of representations as occurring "inside the head." This would surely distinguish thinking as such from experience more generally if, as proposed above, the latter were regarded as typically being extended outside of the head. Thinking would thus be distinguishable as a kind or aspect of experience that essentially takes place entirely in the brain. Stopping to think would then be a matter of "turning inward."

This might serve to "save the appearances," but it is not the tack taken by Mead and Dewey. A pressing problem for Mead and Dewey was to find an account of mind the origins and character of which could be explained in naturalistic evolutionary terms. Mead in particular thought this problem through in some detail and developed a view of thinking as a distinctive kind of extended experience specifically suited to representation and the use of representations. There is no reason to locate this particular kind of activity inside the brain exclusively so long as it can function as a manner of disengagement from automatic activities in order to facilitate the consideration of possible options and the planning of how better to proceed (typically in response to difficulties emerging in merely automatic activities). What distinguishes thinking in this sense from other kinds of "workings of habits" is that it is a kind of reflexivization of the workings of *social* habits. Evolutionarily speaking, we are thinking creatures *because* we are more fundamentally social creatures. In this view, thinking is extended not just because it is a kind of extended experience but also because it specifically involves extensions into social and cultural environments.

We can think only because we can converse, so to speak. How does Mead explain this?

Social and cultural externalism

It has been proposed that a living, growing creature should be conceived as the workings of integrated bundles of habits (routine ways of acting or being, established ways of life, etc.) that as such are not confined in their constitution to the interior of the creature's skin or skull. A habit in itself is to be regarded as a more or less settled (fixed) coordination of bodily and environmental factors. In other words, a live creature is supposed to be a body-in-an-environment whose ways of living are continually evolving in such a manner as to maximize the chances of its staying alive. Ways of living become established (as habits) insofar as they work successfully in promoting survival and well-being. Growth and evolution are matters of the continual development and modification of habits, broadly conceived, such that the latter are precisely what constitute the features and character of a given type of living body-in-an-environment.

Much of the motivation for this way of thinking, for Mead and Dewey, was to avoid the irreparably convoluted inconsistencies of modern epistemology. It called for starting over more or less from scratch to build up a different picture of experience, rationality, and human nature in general. Dewey and Mead consciously sought to begin with *and to maintain at every turn* a conception of the thinking creature as an integrated living body-in-an-environment. Rather than having to bridge an unbridgeable Cartesian chasm between body and mind, the task for Dewey and Mead was to explain how (and what it even means to say that) a species of live creatures might *evolve* mental capabilities.

Burke[19] explains Mead's just-so account of the evolutionary origins of mental capabilities and the capacity for selfhood. Burke and Everett[20] show moreover how Mead's evolutionary social psychology promotes an externalist conception—indeed a socially-externalist conception—of *mind* and *self*. Namely, in line with contemporary cognitive externalism, the world exterior to a given brain and skull includes other brains and skulls, many of which interact in social groups. It is a simple corollary that societies are part of the world external to individual brains and skulls and thus would surely be part of any "extended mind" if anything is.[21] We will review this account in order to better comprehend the role of *representation* in Mead's social psychology.

The following rather dense passage introduces and in effect defines several key terms—"idea," "symbol," "significant symbol," "thought," "thinking," "abstract thinking," "objectivity"—that together give a fairly good picture of Mead's conception of mentality. The terms just listed are italicized in the quote below though not in the original text.

> The possibility of this [adoption of an attitude of the other] entering into his experience we have found in the cortex of the human brain. There the co-ordinations answering to an indefinite number of acts may be excited, and while holding each other in check enter into the neural process of adjustment which leads to the final overt conduct. If one pronounces and hears himself pronounce the word "table," he has aroused in himself the organized attitudes of his response to that object, in the same fashion as that in which he has aroused it in another. We commonly call such an aroused, organized attitude an *idea*, and the ideas of what we are saying accompany all of our significant speech. … Where a vocal gesture uttered by one individual leads to a certain response in another, we may call it a *symbol* of that act; where it arouses in the man who makes it the tendency to the same response, we may call it a *significant symbol*. These organized attitudes which we arouse in ourselves when we talk to others are, then, the ideas which we say are in our minds, and insofar as they arouse the same attitudes in others, they are in their minds, insofar as they are self-conscious in the sense in which I have used that term. But it is not necessary that we should talk to another to have these ideas. We can talk to ourselves, and this we do in the inner[22] forum of what we call *thought*. We are in possession of selves just insofar as we can and do take the attitudes of others toward ourselves and respond to those attitudes. We approve of ourselves and condemn ourselves. We pat ourselves upon the back and in blind fury attack ourselves. We assume the generalized attitude of the group, in the censor that stands at the door of our imagery and inner conversations, and in the affirmation of the laws and axioms of the universe of discourse. … Our *thinking* is an inner conversation in which we may be taking the roles of specific acquaintances over against ourselves, but usually it is with what I have termed the "generalized other" that we converse, and so attain to the levels of *abstract thinking*, and that impersonality, that so-called *objectivity* that we cherish. In this fashion, I conceive, have selves arisen in human behavior and with the selves their minds.[23]

Individuals are able to think, on this account, insofar as they are able to converse reflexively in just the ways that they converse with other social objects, this being possible because those manners of interaction already function habitually if not instinctively in individuals' normal conduct.

To help fill out this story, recall the general sense in which Mead's story is an *evolutionary* story.[24] The basic idea for Mead is that new types of *activities* described at each stage of any such evolutionary account will initially emerge as effective means to achieving various ends (constrained and motivated by changing environmental conditions as much as by established behavioral capabilities). When such activities proved to be valuable (and if they are otherwise naturally selectable), they may generate new species-specific *abilities*. In that case, the possibility of regularly engaging in a given activity becomes an engrained, embodied ability—thus an evolutionary accomplishment, an evolutionary achievement, a further step or stage in evolutionary development of the species.

This suggests a ratcheting process.[25] Evolutionary progress is not just a matter of individuals or species engaging in *new activities*. It also requires stabilizing the capacity to engage in those activities so as to establish new inheritable ready-to-hand *abilities*. It is not just a matter of composition and variation but also of selection and transmission. The combined effects of composition and variation plus selection and transmission of abilities constitute the innovation plus stabilization that characterize a ratcheting process.

It must be emphasized that *abilities*, not genes, are the focal units of evolutionary explanation, both biological and cultural. In particular, the evolution of *social objects* need not be reduced to genetic or genotypic versus phenotypic evolution. Genes are one kind of embodiment of stabilized achievements of new abilities, to whatever extent the latter are biologically transmittable from parent to child. But some abilities—e.g. the corpus of abilities that constitute *language*— are to a great extent transmitted by other means. The Lamarckian notion that "an individual organism changes during its lifetime, and these changes can be passed on to offspring" begins to make some sense here so far as cultural evolution is concerned. The evolutionary stories we have to tell in such cases could *nevertheless* be cast in terms of evolutionary ratcheting processes.

Specifically, a corpus of abilities that constitute *language* has come to be part of the innate hereditary (genetic) legacy of humans, only needing to be triggered and adequately sustained as a child develops. A general capacity is transmitted biologically. Genetic variations may in fact result in certain linguistic *in*capacities,[26] where such exceptions prove the rule, as it were. But specific abilities that constitute competence with particular languages like English or Portuguese are *not* transmitted biologically—English and Portuguese are not hard-wired. These particular abilities are transmitted by *cultural* means that are simply not explicable in terms of genes alone. We can generalize this to virtually any kind of culture-specific abilities.

Mead's story on the whole is a fairly complicated evolutionary story that attempts to explain *how* mental capabilities might have come about at all—as opposed to mere perceptual abilities that would be expected of a healthy frog or any other kind of sentient creature. The story eventually involves the reflexivization of social capabilities in order to produce robustly individual thinking selves.[27] This story is consistent with an externalist view of mind and self beyond the individual skull, not just physically but socially.

The acceptability of Mead's account of human origins hinges particularly on being able to justify the later stages in that account—to show, namely, how the later steps are evolutionarily enforceable without recourse to explanatory skyhooks. The story we want to tell calls for a particular succession of biological ratcheting effects that need not be uniquely human, nor must they have occurred exactly when they did. But as a matter of fact, they developed in such a manner as to enhance human *cultural* ratcheting in unprecedented ways, attaining some kind of threshold momentum by roughly 50,000 years ago. Such evolutionary advances might be the result not just of increasingly complex machinations of symbols and the habits these symbols may trigger but of the emergence of abilities to appreciate and freely manage (1) *recursively enumerable* grammars of symbols that (2) are suited for reference to yet-to-be-actualized possibilities. In this sense, we would have to account not just for the emergence of symbols or even languages, but of symbolic languages that allow open-ended consideration and management of possibilities—thus engendering conditions crucial for the acquisition of essential human characteristics such as the capacity to *think*, *analyze*, and *theorize*. From a functionalist perspective, the thinker's thinking and hence the thinker are to be cast as being constituted by the workings of the abilities that make for such capacities. If parts of these workings take place outside of the skull, then the *thinker* is not to be located wholly within the skull.

It is no surprise of course that Mead's story envisages a kind of social externalism. Mind, as such, is located in a field of conduct that encompasses both individual and object. A pertinent fact here about "an environment" is that some of its constituent objects will in many cases be *social objects*.

What does that mean? In general, established organism/environment relations determine the types of objects that occupy a living being's environment. That is, objects are individuated depending on how the living organism and environment are dynamically related. "Objects exist in nature as patterns of our actions."[28] Different types of living beings may thus have very different environments in one and the same physical expanse.

This relational perspectivity applies all the more to other living creatures of one's kind (or in the group to which the given living creature belongs) such that another of one's kind is a particular sort of object—a *social object*—in one's environment. If Otto's mind encompasses tablets on which he can write, then it just as easily encompasses other people to whom he is appropriately related in some stable fashion and with whom he can converse or otherwise cooperate.[29]

A key component of Mead's story at this point is that one's social group *as a stably organized whole* may be regarded as an agent capable of *gesturing* (that is, as a social object with tangible features). Vocal gestures (as a contingent matter of fact) are especially well suited to facilitating self-consciousness insofar as an individual is easily capable of gesturing to an entire group *and* an entire group may react in kind as a single social object. Vocal-gesturing capabilities of course have a long evolutionary lineage, antedating the emergence of selfhood by a long shot; but that only means that they are present as a reliable medium of gesturing by the time some such medium would figure into the emergence of self-consciousness.[30] The point is that (anticipations of) regularities in individual/group interactions are subject to being incorporated into one's normal (habitual) conduct. An individual, then, will find itself "tending to act toward himself as the other acts toward him" just when (1) vocal gestures made *and heard* by the same individual can initiate tendencies to respond to them as the group would respond to them such that (2) this "beginning of an act of the other in himself" (adopting an attitude of the other, as it were) enters into his own experience. The organized reactions of the group as a whole will thus become "imbedded" in the individual's own conduct. A "generalized other" in effect defines the individual's *character* as a unique personality insofar as it incorporates the group's tendencies into the habitual conduct of the individual.

Complex social institutions can therefore be social objects in this way, namely, where an individual is appropriately related to such institutions in a stable way and with which that individual can dynamically interact. Things like coffee cups, chairs, pens, and paper exhibit various tangible features in a given living creature's experience depending on how they function in the relational dynamics of that experience. Social objects, as physical objects, will likewise have tangible features, and one peculiar class of these tangible features will include so-called "gestures"—a particular kind of stimulus to action that serves as the currency of social interactivity as such.

Another important element of the story here is Mead's conception of an *attitude*. An attitude of a given type (by definition) is a "readiness to act" in some respective way. Adopting an attitude, then, is making ready to act in some respective way, readying respective abilities without completing their execution. In the event of adopting an attitude, the range of possible consequences of executing the respective actions are just that: unactualized possibilities. These possibilities are accessible as objects of attention (available, that is, as *anticipations* or *expectations*) insofar as they are built into the habits that constitute respective abilities that have been readied by adopting the given attitude. Detecting such possibilities in given circumstances is essentially just a case of distance detection, but "distance" in a "space" of possible results of possible courses of action in a quite broad sense. Distance detection, literally, is of course a very primitive capability that emerged very early in animal evolution; so there is nothing magical going on here if such magic was not going on then.

Social acts (whether hostile or benign) typically involve the adoption of preparatory attitudes (posturing in various ways: hailing, greeting, threatening, feigning, placating, etc.) that themselves may be regarded as constituent "utterances" of a sort in a developing conversation of attitudes (for example, consider two grown male dogs sizing each other up, mutually deciding whether or not to fight). Such conversations of *attitudes* can be complicated, and abilities to engage in such complicated conversations can be evolutionarily useful insofar as they introduce a capacity for weighing different possible consequences of various courses of action short of actually executing those actions.

Then, recall Mead's statement quoted earlier that, in more evolved forms, such aroused attitudes are what we would call *ideas*. Ideas in this sense are inherent in the utterances that constitute conversational speech insofar as such utterances are gestures indicating respective attitudes. Somewhere along the evolutionary/developmental line, certain types of gestures will have become associated with respective types of ideas. "Where a vocal gesture uttered by one individual leads to a certain response in another, we may call it a *symbol* of that act; where it arouses in the man who makes it the tendency to the same response, we may call it a *significant symbol*."[31] Such symbolic gestures become the *terms* by means of which *conversations* as social acts may take place.

All of this rests on having stabilized and maintained abilities to converse—abilities that continue to function and change so as to make possible the alteration (degradation or enrichment) of abilities higher (later) in the evolutionary ratcheting hierarchy. Details of how such capabilities come about are sparse, but an account of the emergence of capacities to think pivots on the advent of symbols and symbol systems as means for managing if not exploring different types of attitudes and systems thereof—as means for managing and exploring ideas, as it were. Symbols, recall, are types of gestures, e.g. types of vocal gestures used in conversation or in the coordination of social acts more generally. If we now plug this into the story outlined earlier about a *generalized other* as an interlocutor in reflexive "talk," we get what we could call *thinking* as reflexive talking by means of "significant symbols" (whether silent or audible).

Another important point here—one that shows how Mead's view differs from Hobbesian psychological egoism—is that *social objects* predate *selves*. Our ancestors surely lived as members of well-defined cooperative social groups long before they were selves. Civilization and culture are not the later product of prior full-blown selves eventually deciding by mutual agreement, in the interest of securing peace in a war of all against all, that it would be prudent to bridle and rein in their respective self-serving freedoms. Rather, prior to the evolutionary emergence of *selves* as such, an individual's reactions to the gestures of others, presumably, were to some degree instinctive and automatic (as in sex, parenthood, hostility, etc.). Sociality, sometimes hostile and sometimes not, for better or worse, will have adopted certain interactional and transactional regularities (e.g. hierarchical social structures) that became the stage on which selfhood might have emerged.

Selves emerged in the evolution of the human species (incrementally via natural evolutionary tendencies, just as the development of selfhood in a contemporary human infant is gradual) when individuals became social objects in their own respective experience. This is the case insofar as individuals are capable of *social acts*—"acts which involve the cooperation of more than one individual."[32] One is able to engage in commerce, as a buyer, say, though the part of the exchange of property that belongs to the seller just as necessarily constitutes part of this social act and thus belongs to the buyer as a proper reaction to the buyer's gestures. One typically does not buy and sell to oneself; but an individual can be a social object in its own experience insofar as it adopts the particular role of buyer that requires another individual to play out the complementary role of seller. Both roles in proper relation to one another serve to constitute each individual's habits of conduct even if the individual plays only one of those roles in the execution of a respective social act. An individual is able to be a buyer only because the buyer role has been instantiated in that individual's experience of other social objects so as to provide a template for that particular kind of participation in social acts. Similarly, an individual is able to be a first baseman only because others are in that instance realizing other roles of a baseball game. Then, an individual can be a social object in its own experience (as a buyer or as a first baseman etc.) by virtue of its being the one playing a specific standard role in a standard cooperative act.

Yet another key point is that only selves have minds. Selves predate minds. Selfhood predates mentality. The emergence of mentality is possible only where the emergence of selfhood has laid proper groundwork. How so? Animals will have been sentient at very early evolutionary stages. Mentality (having cognitive capabilities etc.) is more than just sentience. And it is more than just selfhood. Mead takes many pages to develop this part of the story,[33] but a crucial distinction to highlight here is one between becoming a social object in one's own experience (the emergence of selves) and becoming a self in one's own experience (the emergence of "self-consciousness"). An individual appears not just as a *social object* but as a *self* in its own conduct insofar as it is able to take attitudes that others take toward it in these various cooperative interactions.

The notion of *representation* may easily enter the story here insofar as talking and thinking are often addressed to some concrete situation. That is, talking and thinking typically play a role in resolving some given difficulty. We talk to one another often in order to resolve some shared problem. Recall as well that *an experience* or *an act* is an episode of just that—resolving some given difficulty. So talking and thinking—e.g. planning a course of action, or simply exploring possible courses of action—typically play a role in furthering the completion of some act. *An inquiry*, for Dewey, is such an act in which talking and thinking play a crucial role in furthering its completion. This crucial role, in broadest terms, could be said to be representational in character. The respective talking and thinking is *about* facets of the problem at hand, particularly in regard to exploring and assessing its particular features and its potentials for resolution. A plan, after all, is a "representation" of a possible future course of action. This role of thinking in inquiry can be carried out "symbolically" as it were.

Representational externalism, extended computation

On Mead's account, the prior emergence of abilities to *converse* eventually gave rise to abilities to *think*. Abilities to think presuppose abilities to converse. There is nothing in this account to suggest that the workings of symbolic representation that constitute thinking occur only in brains or inside skulls. At least some such workings might occur only by way of *talking*, or by way of *writing* things out and exploring their implications on paper, chalkboard, etc. A pragmatist account of cognitive externalism[34] is quite general but shows in particular how the (fast) workings of perception (directly accessing information outside of the head) are extended. The (slow) workings of representation are also extended to the extent that their symbolic media are extended. These two kinds of workings will *correspond* in some functional way insofar as talking and thinking registers and assesses possibilities while perceiving accesses and classifies present actualities,[35] each pertaining to a common inquiry (a common experience, a common act).

This yields a type of representational and/or computational externalism—a theory, more or less, of extended computation and/or representation. Recall that the workings of habits that constitute an *experiencer* (not necessarily a thinker) are easily conceived as extended across many causally coupled things and activities inside and outside of the head. This view was characterized earlier as follows in the chapter on page 182.

Notice that the first five of these six characteristics, while essential to thinking, do not require thinking; and the sixth characteristic only suggests how we might begin to distinguish thinking from other elements of experience.

Consider the following example. The antisnake behaviors of California ground squirrels are complicated, involving intricately complex sequences of choice points and behavioral transitions that in any case do not require thinking as such.[36] An internalist conception of experience is not *needed* to account for these behaviors. An externalist conception of experience may do just as well. More importantly, if we consider what it would mean for a squirrel to *think* about what it is doing, we still need not introduce an internalist viewpoint.[37]

Given that none of the points (1)–(5) requires an internalist framework, it is interesting to speculate about what it would take for such a creature to engage in or to evolve capabilities of thinking. In line with point (2) above, if we regard the squirrel's brain neither as its entire CPU nor as its sole "memory" bank but rather as a chipset with the job of mediating extended organism/environment life-activities, then we cannot appeal to mere "complexity" of habits to account for what is rather a matter of distinguishing functional roles. That is, it would beg the question at hand to imagine that a *thinking* squirrel would simply be one with a more complex brain solely capable of supporting, say, internal representational activities (the latter being thinking activities).

An externalist can agree that thinking is essentially representational in character, and this may require a more complex chipset inside the head, but it does not require that representational activities as such take place exclusively inside the head. What distinguishes representational activities as thinking activities is not where they occur so much as how they function throughout the course of an experience or act to facilitate achievement of the purpose at hand. Evidence that a squirrel is able to think, on this account, would probably involve activities like rehearsing or planning or practicing search-and-destroy procedures—where acting out such procedures in the absence of any evidence of snakes would be a way of *representing* what may transpire when a snake is encountered. The same behaviors that occur instinctively in the presence of snakes would now presumably, in the absence of snakes, play a different functional role (as rehearsing or planning or practicing or whatever the case may be). Granted, it is not easy to imagine a real squirrel engaging in these kinds of representational behaviors, nor is it at all clear how or why such representational capabilities would ever have evolved as part of a squirrel's innate repertoire of habits (namely, thinking can often slow things down, which is not so good when facing a rattlesnake ready to strike), though it does help to clarify what an externalist conception of thinking might look like.

It is easier to imagine such representational behaviors taking place in human experiences—where a different evolutionary groundwork is in place. Human behaviors include the use of symbols and symbolic media. For example, planning next year's garden requires some thinking. Such thinking will proceed as one draws up lists, charts, diagrams, tables, calendars, etc. representing on paper, say, what may actually transpire next year. The nice thing about pen and paper is that, with sufficient care, information etched onto the paper will not easily degrade and thus may be repeatedly accessed, modified, and otherwise utilized. This kind of information storage does not take place solely within a normal brain. *Something* takes place solely within the normal brain throughout such a thinking process, but that something typically is not all there is to thinking or even representation—just as what happens in the chipset of a computer is not all there is to computing.

Likewise, a theoretical physicist *A* may work for some time with chalk and chalkboard, or with pencil and paper, working out the mathematical details of some particular model of a certain kind of physical system that so far is not well understood. Tweaking the model in various ways will of course draw out the process in more extensive and complicated ways. These representational behaviors— *A*'s thinking—will encompass the making of marks on paper or chalkboard as much as anything else. In some cases such representational behaviors may take place only in *A*'s head, but we would want to say that, typically, thinking could not take place in any sustainable way without such external markings to the extent that such markings are a constituent part of the workings of representation and thus *memory*. Physicist *A* may *utter the results* of such cogitations by means of a written publication in an appropriate journal, or perhaps more simply by way of direct conversation with a colleague *B* who is able to see to testing those results in a laboratory setting—able to perform activities whose actual results should but may not correspond to what *A*'s cogitations will have predicted in such circumstances.

Such examples exhibit the perspective one must take to properly characterize mental processes. Not having the thinker's present purposes in view would be like assessing a painting by analyzing the chemistry of the pigments, or judging a novel by analyzing the grammar, or worse, only the font that was used in the printing. Not taking into account different functional roles that otherwise similar behaviors may play in pursuing such purposes leaves only bodily motions as the subject-matter of behavioral psychology.

The generic template for an act (an experience, an inquiry) consists then of two types of extended "workings" in the development of the act: one that tracks actual facts of the case, and one that represents such facts and their implications in order to track possible courses of further development.[38] The common purpose for these two types of workings is the completion of the given act, namely, effective resolution of the difficulty or disturbance that initiated the act. Generally speaking, sources of difficulties would include almost anything that occurs in the life of an experiencer, not excluding the very coordination (correspondence, mutual coherence) of these two basic types of workings insofar as they constitute the life of a *thinking* experiencer.

Illustrations of the latter would include numerous episodes in the history of any given science. Such episodes emerge and develop and are resolved or not by way of the workings of both theoretical and experimental activities in coordination with one another (often resulting in their mutual reconstruction by way of modified theoretical and/or experimental techniques and instrumentation). The extended nature of both of these types of workings in science is obvious. Experimentation consists of fact-finding activities that include manipulations of environmental circumstances through the controlled use of appropriate conditioning and measuring instruments. Theoretical activities consist of idea-finding activities that include manipulations of public discourse that may involve alterations in linguistic and analytical methods.

It is crucial to note, of course, that the extended nature of this generic template of coordinated fact-finding and idea-finding is exemplified in the acts/inquiries of any single thinking experiencer. Instances of silent and dark reflexive discourse employ symbols and techniques derived from ever-evolving languages and conversations in one's community, and one could argue that only such silent and non-visual reflexive discourse counts as thinking and that it may easily be regarded as an activity taking place only inside the skull. On the contrary, such discourse, while occurring not infrequently, is perhaps less common than might be expected. Such reflexive discourse (silent and dark or not) utilizes a language of some respective community and will sooner or later have to be represented in some sensible if not public way just to recalibrate its terminology and semantics, given (1) that the public language from which the language of one's private thinking is derived may be assumed to be always changing (so that remaining proficient in using such a language is like trying to hit a moving target), and (2) that one is no doubt always continuing to *learn* the language to a greater extent or in finer detail, which necessarily involves direct participation in its public use.

Basically, one uses a language privately (silently etc.) not unlike the way one utilizes air that has been inhaled—internally, inside the lungs—such that one must sooner rather than later exhale. The content of the language "inhaled" into one's cortex or wherever is only as good as the scope and detail of the generality of the *generalized other* with whom one silently and darkly converses, and the continual growth and development of that generalized other as a consequence of ongoing interaction with particular others other than oneself is surely a major part of the extended workings that constitute the thinker.

Notes

1 This is the classical view promoted by Allen Newell and Herbert Simon 1976, pp. 113–26.

2 For instance, see Robert Shaw 2003, pp. 37–106; Robert Port and Timothy van Gelder 1995; Rodney Brooks 1991, pp. 139–59; Scott Kelso 1995.

3 See Frederick Adams and Kenneth Aizawa 2001, pp. 43–64; Frederick Adams and Kenneth Aizawa 2008; Richard Menary 2010a; F. Thomas Burke and Stephen Everett, 2013, pp. 107–35.

4 For example, see John Dewey 1925, 1938, 1912, pp. 401–6, 1924–5, pp. 251–77, and 1938, ch. 21, pp. 357–442.

5 The term "chipset" is ambiguous. Here it means a particular group of chips on a motherboard that controls many of the latter's data-transfer capabilities (by way of both high-speed and low-speed channels), providing in effect the interfaces among a computer's subsystems—thus mediating interactions among, for example, the CPU, different kinds of so-called memory, and I/O devices.

6 Richard Menary 2010b, pp. 459–63.

7 Ecological psychology emerged in the 1940s from military research into the nature of perceiving moving objects from moving perspectives, e.g. sighting and identifying an airplane in flight from the point of view of the pilot of another

airplane in flight (Gibson 1950). This research was largely eclipsed by GOFAIR in later decades, though it continues to have a tenacious following and is responsible for some of the recent innovations in the cognitive sciences.

8 Andy Clark and David Chalmers 1998, pp. 10–23.

9 Hilary Putnam 1960, pp. 148–80.

10 Gibson's notion of a perceptual system being *attuned* to information in the world may be useful here. As a matter simply of terminology, we could say that the difference between an organism's *environment* (those parts of the world to which an organism's perceptual systems are attuned) and the *world at large* is that the former includes only what its perceptual systems are capable of accessing immediately. The organism's perceptual systems, in such a picture, would extend into the world at large to include its environment. Creatures capable of perceiving should thus by default be regarded as *integrated* body/environment systems, to use Menary's term. As stated, this does not identify "perceiving" with "accessing information immediately." It only says that information that is accessed and utilized to constitute *perception* may be immediately available in the world outside of the skull and thus need not be "represented" internally in order to be accessed and utilized. Representations may be necessary for *thought*, but not for perception.

11 John Dewey 1934.

12 George Herbert Mead 1938a and 1907, pp. 383–90.

13 John Dewey 1896, pp. 357–70.

14 Dewey 1938.

15 Charles Sanders Peirce 1877, pp. 1–15, and 1878, pp. 286–302.

16 Peirce 1878.

17 Adams and Aizawa 2001 and 2008.

18 The issue here is mind, mentality, thinking. This is not the problem of consciousness. That problem is already present as a mystery in an account of experience as extended, in the form of what we might refer to as the problem of *sentience*. It is assumed here that animals other than humans are sentient creatures. They are experiencers. They are capable of feeling pain etc. If we could explain the nature of that feeling, then we can very likely explain consciousness as a variation on that theme.

19 Tom Burke 2005, pp. 567–601.

20 Burke and Everett 2013.

21 For instance, see David Spurrett and Stephen Cowley 2010, pp. 295–324.

22 It is unfortunate that Mead should use the term "inner" here, unless it refers to being inside the self but not necessarily inside the skull. The term "reflexive" is a better term. As with arithmetic calculations solely in our heads or with the aid of pencil and paper, whether or not talking to oneself takes place in a forum of thought is not decided by whether it proceeds silently or out loud.

23 Mead 1924–5, pp. 287–8.
24 Burke 2005, pp. 571–3.
25 Michael Tomasello et al. 1993, pp. 495–552; Michael Tomasello 1995, pp. 131–56, and 1999.
26 Cecelia Lai et al. 2001, pp. 519–23; Wolfgang Enard et al. 2002, pp. 869–72.
27 The main elements of that account can be gleaned from George Herbert Mead 1956 and 1964. See also Hans Joas 1985, ch. 5.
28 Mead 1924–5, p. 289.
29 Clark and Chalmers 1998.
30 Robin Dunbar 1997.
31 Mead 1924–5, pp. 287–8.
32 Mead 1924–5, p. 279.
33 For example, Mead 1924–5, pp. 278–89.
34 Burke and Everett 2013.
35 This does not mean that perceiving (e.g. perceiving distances) is not prospective. But such instinctive anticipations that result from perceiving are typically not the subject of thought, that is, unless they are or become in some way relevant to what is being thought about. It may sound trivial to put it that way, but the point is that possibilities that are occurrent as such in perception are not for that reason alone the subject of thought. That will depend on what their role or function becomes in the given inquiry. Similar ideas are tentatively explored in Burke 2008 and more fully in 2013, ch. 7.
36 Richard Coss 1989, 1991, pp. 277–315, and 1993, pp. 153–94; Richard Coss and Donald Owings 1989, pp. 30–5; Matthew Rowe and Donald Owings 1989; Gregory Burton 1993, pp. 153–69; Donald Owings 2002, pp. 19–25.
37 Burke and Everett 2013.
38 Burke 2008.

References

Adams, Frederick and Aizawa, Kenneth (2001), "The Bounds of Cognition," *Philosophical Psychology* 14: 43–64.

— (2008), *The Bounds of Cognition*. Malden, MA: Blackwell Publishing.

Brooks, Rodney A. (1991), "Intelligence without Representation," *Artificial Intelligence* 47: 139–59.

Burke, F. Thomas (2005), "The Role of Abstract Reference in Mead's Account of Human Origins," *Transactions of the Charles S. Peirce Society* 41(3): 567–601.

—(2008), "(Anti)Realist Implications of Pragmatism's Dual-Process Active-Externalist Theory of Experience," *Philosophia Scientiae* 12(1): 187–211.

— (2013), *What Pragmatism Was*. American Philosophy Series. Indianapolis, IN: Indiana University Press.

Burke, F. Thomas and Everett, Stephen (2013), "Social-Psychological Externalism and the Coupling-Constitution Fallacy," in *George Herbert Mead in the Twenty-First Century*, F. Thomas Burke and Krzysztof Piotr Skowroński (eds). Lanham, MD: Lexington Books, pp. 107–35.

Burton, Gregory (1993), "Behavioral Relics and Animal-Environment Mutualism: Commentary on Coss 'Context and Animal Behavior III' (1991)," *Ecological Psychology* 5(2): 153–69.

Clark, Andy and Chalmers, David (1998), "The Extended Mind," *Analysis* 58: 10–23.

Coss, Richard G. (1989), "Evolutionary Persistence and the Contextual Modulation of Antisnake Behavior." Presented at the Fifth International Conference on Event Perception and Action. Oxford, OH, July 1989.

—(1991), "Context and Animal Behavior III: The relationship between early development and evolutionary persistence of ground squirrel antisnake behavior," *Ecological Psychology* 3(4): 277–315.

—(1993), "Evolutionary Persistence of Ground Squirrel Antisnake Behavior: Reflections on Burton's commentary," *Ecological Psychology* 5(2): 153–94.

Coss, Richard G., and Owings, Donald H. (1989), "Rattler Battlers," *Natural History* 98(5): 30–5.

Dewey, John (1896), "The Reflex Arc Concept in Psychology," *Psychological Review* 3: 357–70. Reprinted in *Philosophy and Civilization*, 233–48, and in *EW5*: 96–110.

—(1925), *Experience and Nature*. Chicago, IL: Open Court. Reprinted in *LW1*.

—(1931), *Philosophy and Civilization*. New York: Minton, Balch.

—(1934), *Art as Experience*. New York: Henry Holt. Reprinted in *LW10*.

—(1938), *Logic: The Theory of Inquiry*. New York: Henry Holt. Reprinted in *LW12*.

—(1967–72), *The Early Works, vols 1–5 (1882–1898)*, ed. Jo Ann Boydston. Carbondale, IL: Southern Illinois University Press. Citations of items in this edition are indicated by EW followed by volume and page numbers.

—(1976–80), *The Middle Works, vols 1–15 (1899–1924)*, ed. Jo Ann Boydston. Carbondale, IL: Southern Illinois University Press. Citations of items in this edition are indicated by MW followed by volume and page numbers.

—(1981–90), *The Later Works, vols 1–17 (1925–1953)*, ed. Jo Ann Boydston. Carbondale, IL: Southern Illinois University Press. Citations of items in this edition are indicated by LW followed by volume and page numbers.

Dunbar, Robin I. M. (1997), *Grooming, Gossip, and the Evolution of Language*. Cambridge, MA: Harvard University Press.

Enard, Wolfgang, Przeworski, Molly, Fisher, Simon E., Lai, Cecelia S. L., Wiebe, Victor, Kitano, Takashi, Monaco, Anthony P., and Pääbo, Svante (2002), "Molecular Evolution of FOXP2, a Gene Involved in Speech and Language," *Nature* 418: 869–72.

Everett, Stephen (2013), *Vehicle Externalism, the Coupling-Constitution Fallacy, and Dewey*. PhD thesis. University of South Carolina.

Gibson, James J. (1950), *The Perception of the Visual World*. Boston, Houghton Mifflin Company.
Joas, Hans (1985), *G. H. Mead: A Contemporary Re-examination of His Thought*. Trans. Raymond Meyer. Cambridge, MA: MIT Press. Originally published as *Praktische Intersubjektivität, Die Entwicklung des Werkes von George Herbert Mead*. Frankfurt, Suhrkamp-Verlag, 1980.
Kelso, J. A. Scott (1995), *Dynamic Patterns: The Self-Organization of Brain and Behavior*. Cambridge, MA: MIT Press.
Lai, Cecelia S. L., Fisher, Simon E., Hurst, Jane A., Vargha-Khadem, Faraneh, and Monaco, Anthony P. (2001), "A Forkhead-Domain Gene is Mutated in a Severe Speech and Language Disorder," *Nature* 413: 519–23.
Mead, George Herbert (1907), "Concerning Animal Perception," *Psychological Review* 14: 383–90. Reprinted in Mead 1964, 73–81.
—(1912), "The Mechanism of Social Consciousness," *Journal of Philosophy, Psychology, and Scientific Methods* 9: 401–6. Reprinted in Mead 1964, 134–41.
—(1924–25), "The Genesis of the Self and Social Control," *International Journal of Ethics* 35: 251–77. Reprinted in Mead 1932, 176–95, and in Mead 1964, 267–93.
—(1932), *The Philosophy of the Present*, ed. Arthur E. Murphy. La Salle, IL: Open Court.
—(1938a), *The Philosophy of the Act*, ed. Charles W. Morris. Chicago, IL: University of Chicago Press.
—(1938b), "The Process of Mind in Nature," in *The Philosophy of the Act*, ed. Charles W. Morris. Chicago, IL: University of Chicago Press, ch. 21, pp. 357–442. Reprinted in Mead 1956, 85–111.
—(1956), *On Social Psychology*, ed. Anselm Strauss. Chicago, IL: University of Chicago Press.
—(1964), *Selected Writings*, ed. Andrew J. Reck. Chicago, IL: University of Chicago Press.
Menary, Richard, ed. (2010a), *The Extended Mind*. Cambridge, MA: MIT Press.
—(2010b), "Introduction to the Special Issue on 4E Cognition," *Phenomenology and the Cognitive Sciences* 9: 459–63.
—(2012), "Cognitive Practices and Cognitive Character," *Philosophical Explorations* 15(2): 147–64.
Newell, Allen and Simon, Herbert A. (1976), "Computer Science as Empirical Inquiry: Symbols and Search," *Communications of the ACM* 19(3): 113–26.
Owings, Donald H. (2002), "The Cognitive Defender: How Ground Squirrels Assess Their Predators," in *The Cognitive Animal: Empirical and Theoretical Perspectives on Animal Cognition*, Marc Bekoff, Colin Allen, and Gordon M. Burghardt (eds). Cambridge, MA: MIT Press, pp. 19–25.
Peirce, Charles Sanders (1877), "The Fixation of Belief," *Popular Science Monthly* 12: 1–15. Reprinted in *The Essential Peirce*, vol. 1, ch. 7.
—(1878), "How to Make Our Ideas Clear," *Popular Science Monthly* 12: 286–302. Reprinted in *The Essential Peirce*, vol. 1, ch. 8.

—(1992/1998), *The Essential Peirce: Selected Philosophical Writings*, in two volumes, ed. Peirce Edition Project. Bloomington and Indianapolis, IN: Indiana University Press.

Port, Robert and van Gelder, Timothy, eds (1995), *Mind as Motion: Explorations in the Dynamics of Cognition*. Cambridge, MA: MIT Press.

Putnam, Hilary (1960), "Minds and Machines," in *Dimensions of Mind: A Symposium*, ed. Sidney Hook. New York: New York University Press, pp. 148–80.

Rowe, Matthew P., and Owings, Donald H. (1989), "The Information Afforded by Rattlesnake Rattles: A Study of Risk Assessment by California Ground Squirrels." Presented at the Fifth International Conference on Event Perception and Action. Oxford, OH, July 1989.

Shaw, Robert E. (2003), "The Agent–Environment Interface: Simon's indirect or Gibson's direct coupling?" *Ecological Psychology* 15(1): 37–106.

Spurrett, David and Cowley, Stephen (2010), "The Extended Infant: Utterance-Activity and Distributed Cognition," in *The Extended Mind*, ed. Richard Menary. Cambridge, MA: MIT Press, pp. 295–324.

Tomasello, Michael (1995), "Language is Not an Instinct," *Cognitive Development* 10: 131–56.

—(1999), *Cultural Origins of Human Cognition*. Cambridge, MA: Harvard University Press.

Tomasello, Michael, Kruger, A. C., and Ratner, H. H. (1993), "Cultural Learning," *Behavioral and Brain Sciences* 16: 495–552.

10

The Self as an Evolved Organism that Lives in a Pragmatically Defined World

David L. Thompson

Introduction

Pragmatism, as I use the term, is an account of the nature of objects. Realism holds that there is a world of objects that are what they are in-themselves. Idealism claims that objects are what they are only because of mind, subjectivity, or consciousness; in the extreme version, objects are ideas in the mind. Pragmatism, in the philosophical sense, claims that objects get their "whatness" from action—"praxis" is Greek for action.[1] These are, of course, caricatures: no actual philosopher says anything that simple. Still, they map out the terrain for the topics I wish to discuss.

Traditionally pragmatism claims that the essence or meaning of objects depends on human needs and activities. The central argument of this chapter is that we should expand this human-centered account to the action of organisms in general. In the first part, I will argue that biological life involves the evolution of systems that constitute a world of objects in correlation with organisms. In the second part, I will sketch out how we might apply this scheme of thought to human life, to selfhood, and show that the structure within which persons exist is similar to and continuous with the systems of organic life.

Organic life

Emergent worlds

Toy amoeba

Let me start with a hypothetical "just-so" story about the evolution of life. Initially, when life was appearing, complex molecules were governed solely by direct physical and chemical causation. With the development of cells, some substances in the environment—perhaps methane—caused single-cell bacteria to grow and multiply, while other substances—perhaps oxygen—caused them to die. Later some organisms developed a more sophisticated pattern of interaction with their environment. Instead of receptors for each specific chemical stimulus having its own causal response, a structure evolved in which multiple receptors for different chemicals were linked to the same response mechanism. Let me call a unified cluster of such receptors a "detector." A detector enables an organism to react in the same way to different physical stimuli. If these different inputs all contributed to the growth of the cell, then the single, unified response mechanism would presumably be more efficient. In that case, it would be an evolutionary adaptation and would be selected in competition with the more primitive structure of separate response mechanisms for each receptor. What unifies the cluster of receptors into a "detector" is the single response mechanism to which they are connected.

Less abstractly, imagine a single-celled organism, let me call it an amoeba, that gets its energy by absorbing other organisms, say, bacteria, plankton, and diatoms. It detects such energy sources by having receptors that react to different chemicals—scents—given off by these various energy sources. In the evolutionary adaptation I'm proposing, these various inputs are unified into a "detector" that then releases the mechanism for the approach response. I can now describe the situation more elegantly: the amoeba comes to treat these various structures in the environment as the same kind of entity—"food-for-amoeba." Similarly it may treat various other substances, such as iron, ammonia, or alcohol as "poisons to be avoided." Since developing detectors for neutral substances, such as silicon dioxide, would be a waste of biological energy, it is unlikely such detectors would evolve. After many generations of evolutionary selection, most surviving amoebas will have such unified detectors linked to their appropriate response mechanisms.

My scenario, imaginary and oversimplified though it is, can serve as a toy world for us to play with and clarify some of the most fundamental concepts of life: the emergence of organisms; generality and the categorization of objects; functionality and normativity; and the formation of a niche or world. I will discuss each of these four in turn, and then consider some misinterpretations of the development.

Emergent organisms

First, the evolution of the new structure is the emergence of a new unity. When an amoeba's receptor for plankton scent encounters a chemical that stimulates it, what happens next does not depend just on the receptor. What behavior an input releases will no longer simply be the direct, linear effect of the stimulus as cause: the organism processes the input in such a way that the output accomplishes what the organism needs. It is no longer simply the plaything of brute physical causes. The amoeba acts as a whole. As Merleau-Ponty puts it, "Between the exterior world and the living organism, there is an insertion of a whole that orders, coordinates and interprets ..."[2]

To this extent, the organism is partly independent of its causal environment. The organism has become its own regulator: causes are not allowed to automatically unleash their effects. The movement of a dust particle on a stream of water is the direct effect of the force of the stream as cause. But in the case of my amoeba, the force of the stream does not necessarily carry the organism along with it for, if it carries the scent of plankton that the amoeba needs to engulf, the final movement cannot be predicted solely on the basis of the water flow as cause. The stimulus is assigned a role, so to understand the effect of the stimulus we must take into account the internal structure of the organism as well as the nature of the input. The amoeba can be said to "act;" it is (in a very weak sense) an "agent." That's what it is to be a unified, independent, organized organism.

The new unity also involves the emergence of a supervenient regularity; let me call it the Amoeba Food-approach Law. I call it supervenient to make it clear that there is nothing mystical about this emergence. The new regularity arises from the subvenient physical laws, which lose none of their validity in the new situation. Indeed in any *token* case of the application of the Food-approach Law, for example, approaching this particular piece of plankton, we can explain what happens in terms of the underlying physical laws. What the physical laws cannot explain is the *type*, the repetitive pattern. The supervenient, Food-approach Law picks out a regularity of behavior to which the physical laws are blind.

I call it a law because, unlike a mere description, it has explanatory power. A type-based law differs from a superficial collection or redescription of token events in its counterfactual force. The Food-approach Law not only explains those events which actually happen; it also allows us to predict what *would* happen if, contrary to the facts, silicon dioxide could be digested by the amoeba. Once we understand that it is evolutionary fitness that brought about the law in the first place, we can see that silicon dioxide would likely have been included in the unified detector mechanism for food if there would have been a selective advantage in doing so. The Food-approach Law is therefore capable of explaining individual events such as why, in a particular case, this amoeba approaches that diatom as opposed to avoiding it.

The Food-approach Law, however, is a truly emergent law in the sense that it cannot be reduced to the subvenient physical processes: the law only emerges within the contingent, biological history of the Earth during its first few billion years. If history had been different, as it appears to have been on Mars, then an Amoeba Food-approach Law would not have come into being. Since its emergence depends on the contingent particular, historical conditions on early Earth, it could not be predicted by, nor reduced to, physical laws alone. I will elaborate on the role of history later.

Generality and object

Secondly, we can say that the unification of the amoeba's detectors is the "generalization" of its stimulus. That is, a particular piece of plankton may cause the amoeba to approach it, but it does so because that token piece of plankton is treated as a particular instance of the general type "food-for-amoeba." Any substance that qualifies as food-for-amoeba would have the same effect. Substances, however different they might be in their physical or chemical composition, are treated by the amoeba as if they were the same; they are given a status as instantiations of the one generality, food-for-amoeba. This general classification has come into being because of the internal structure, needs, and actions of the amoeba. It is a pragmatic category, in contrast with the categories of physics or chemistry.

Brandom calls this approach a pragmatic version of classification and traces the notion back to Hegel:

> … if it is de-intellectualized, stripped of residual commitments to understanding concepts as explicit to the mind … the roots of conceptual classification are to be found in treating something in practice as being of a certain kind—taking something [particular] as something [universal], by behaving towards it in a way that assimilates it to others. Particular objects are classified as belonging together in some respect by being responded to alike in practice. A respect of similarity in what is responded to then corresponds to a repeatable response. Hegel develops such a pragmatic, indeed naturalized, version of Kant's account in the form of an erotic theory of the origin of awareness, an account of animal desire as the source of classification. As he puts it, an animal classifies some

particular as food when it "falls to without further ado and eats it up." Eating something is treating it, responding to it, classifying it in practice as food.[3]

It is tempting to say that the amoeba now operates with the general concept, "food-for-amoeba," but we need two caveats. First, we tend to associate "concept" with a conscious conceiver, but it would be absurd to suggest that an amoeba is conscious. It is not that the amoeba "thinks" of substances as food; amoebas don't think! It is in its actions that the amoeba treats them as the same. So it is only if we dissociate the notion of consciousness from "concept"—what Brandom calls "de-intellectualize"—that the word might be appropriate to our context. Secondly, we must remember that food is a category of objects in the world; in calling food a concept we must not interpret it as an idea internal to an amoebic mind. An amoeba has no mind! Only with these provisos can we talk of food-for-amoeba as a "concept." Given these associations with the word, it is safer to avoid the loaded term "concept" and to simply say that the actions of the amoeba pragmatically categorize its world.

To put the same point another way, we can say that the development of the detectors in the amoeba leads to the appearance of a new *object* in its world: food-for-amoeba. Where before there were various bacteria, plankton, algae, and so on, all physically and chemically diverse, there is now a new kind of object.

What do I mean by an "object?" We call something an object insofar as it is an enduring unity of a particular kind. As a unity, it must be bounded: it includes parts of the spatio-temporal world, and excludes other parts. For instance, a tree has spatial boundaries that exclude the surrounding air and soil. It also has temporal boundaries from the time of the seed to the old tree's decay. An object must endure for some period so that it can participate in different events as the selfsame entity. These boundaries make it an individual that is differentiated from other individual objects. It is this tree, not that tree. The individual, however, must also be of a kind. This individual is of the same kind as other trees. Whether in fact other individuals of the same kind actually exist or not, it is essential that they *could* exist. It is the general characteristics of any tree that account for the boundaries that individuate this particular tree: it is because the individual is of the kind tree that the soil, air, and original seed are outside its boundary. Finally, an object interacts with other objects as a unity: it can be said to be a cause or effect of events.

It is in this sense that food-for-amoeba is a new, emergent object. In pragmatically allocating the various substances that serve as food to the one category, the amoeba sets the criteria for differentiating food from, say, poison-for-amoeba. It is also the general category that unifies and sets the boundaries for individual instances of food. The boundaries that differentiate food from the surrounding water or rocks are pragmatically determined by the needs of the amoeba. If it could also eat the rocks the bacterium clings to, they would together form just one entity in the amoeba's world. That is to say, food-for-amoeba is a constructed reality: the objecthood of food-for-amoeba is constituted by the same process that, within the amoeba, unifies the latter as a single perceiving, acting organism.[4]

Function and norm

Thirdly, our toy world allows us to explore the genesis of function and norm. The food detector doesn't simply, as a matter of *fact*, detect food; its *function* is to detect food. How do we get from fact to function? Let me offer a historical analysis. In the first amoeba that developed a unified detector, perhaps by genetic mutation, the internal structure just happened by chance to respond to the class of various substances that the amoeba could feed on. This was just a fact, a random event; it was not yet the function of this structure. After many generations, when later amoebas inherit this capacity precisely because their ancestors replicated only because they had this new ability, the structure takes on the function of being a food detector. This becomes its goal or *telos*. Functionalization must be understood as a historical process. To refer to the "function" of a structure is shorthand for the long evolutionary narrative that explains its presence. That a detector is detecting food is a current, momentary fact; to say it has the function of detecting food is to place it in the temporal context of its evolutionary genesis. Functionality is a sedimentation of the past, a congealed history.

The unity of the amoeba itself is a functional unity. It is not just that its parts, its detectors, response mechanisms, and so on, are in a close spatial arrangement within a boundary, within the membrane, it is that each component process, whether provoked by external objects or by something internal, contributes to the overall life of the organism as a whole.

Accordingly, the objects constituted in its world, food-for-amoeba, poison, and so on, are functional objects whose essence depends on their roles within the life of the organism. Kant, over two centuries ago, already put the point elegantly:

> It is no doubt the case that in an animal body, for example, many parts might be explained as accretions on simple mechanical laws (as skin, bone, hair). Yet

> the cause that accumulates the appropriate material, modifies and fashions it, and deposits it in its proper place, must always be judged teleologically. Hence, everything in the body must be regarded as organized, and everything, also, in a certain relation to the thing is itself in turn an organ.[5]

A function is multiply realizable, hence its definition has a certain independence from its material substrate. This is why the chemical structure of the different substances that count as food-for-amoeba is secondary; any one of a number of chemical compounds would do, provided only that they are such that the amoeba can detect and engulf them. My claim above that the Amoeba Food-approach Law has counterfactual force is equivalent to the claim that as a functional regularity it is realizable in an indefinite number of ways. It is not the material composition that determines the function; rather it is the functional role that determines which material substrates will qualify as objects of the kind "food-for-amoeba."

Actions too are functional unities. If we are careful to dissociate the concept of action from explicit, conscious goal-setting—which the amoeba obviously does not do—we can use the term to distinguish active movements attributable to the amoeba from those that are just the passive effects of external forces. An action is a set of physical movements that have a function and can be attributed to the amoeba as a whole. For instance, it is the functional significance of actively moving to engulf food that distinguishes it as an action from the passive movement of being carried along by a stream of water. What defines a particular action—which action it is—depends on the role it plays in the life of the organism. The amoeba's action is not "moving toward a bacterium;" it is "pursuing food-for-amoeba." Like objects, actions are general, that is, they are functionally defined. They are multiply realizable: different physical movements can constitute the same action. So if we leave aside intellectualist interpretations and anthropomorphic associations, we can say that, as a functional unity, the amoeba is an agent: it acts.

The arrival of functionality is also the arrival of normativity. Only where a process has a function can it be said to be correct or incorrect, right or wrong. Once the function of a structure within the amoeba is to detect food, then its positive response to a poison such as ammonia, is a malfunction, a mistake, an error, not just a neutral happening. Actions also are normative: like all functional entities, they can succeed or fail. Viewed functionally, a correct response is no longer just a factual reaction to the present stimulus; it is what this organ *should* do. As a sedimented embodiment of the history of the species the food detector has a purpose, a *telos*, that sets a norm against which an individual event can be judged to be a success or a failure. Functional objects are value-laden. Food-for-amoeba is an object that is *worth* engulfing; the amoeba *should* approach and engulf it, whether in fact it does so or not. This normativity is invisible if we look exclusively at the present event from the viewpoint of physics: only against the context of evolutionary history does its functional normativity become visible.

Already with our simple amoeba we have moved from the purely physical world of facts to a world at least partly defined by value. As Merleau-Ponty puts it, "Correlation by laws, as the mode of thinking in physics practices it, leaves a residue in the phenomena of life which is accessible to another kind of correlation: coordination by meaning."[6]

World-for-the-organism

Fourthly, my toy amoeba scenario lets us explore the distinction between a world-in-itself and a world-for-an-organism. The amoeba "fits" its niche. This is not an empirical fact we might discover by empirical investigation; it is but another way of describing the organism–object relationship. Some have expressed surprise, even wonder, at the "miraculous" fit between an organism and its environment. Fodor has responded that this is a "rigged game:" "… if a creature fails to fit an ecological niche exactly, it follows that that isn't exactly the creature's ecological niche."[7] That an organism fit its niche, he claims, is a tautology. Fish are not fit for life in the trees around the lake nor are many birds fit for swimming in the deeps. The fit appears only when we relate the organism to the specific aspects of the environment relevant to the species.

> Here's the point: a creature's ecology must not be confused with its environment. The environment that creatures live in is common to each and every one of them – it's just "the world" … By contrast, a creature's ecology consists of whatever-it-is-about-the-world that makes its phenotype viable. That's to say: it is constituted of those features of the world in virtue of which that kind of creature is able to make a living in the world. In effect, the notions "ecology" and "phenotype" (unlike the notions "environment" and "phenotype") are interdefined.[8]

Fodor's response is helpful, but not quite adequate. His phrase "the [one and only?] world" suggests that he is talking of the world as defined by physical categories, and that it is a selection of things from this world that makes up the ecology of the species. From my perspective, this misses the crucial point. An organism does not relate to objects defined by physical terms, however carefully we select them for relevance. An organism's niche—its "ecology"—is composed of functional objects constituted by the actions and perceptual system of the organism. The organism and its niche are two sides of the same coin. So fitness is no miracle: it is inherent in the pragmatic nature of objects-for-an-organism.

Uexküll puts it better: each organism lives in its own environment [*umwelt*], made up of objects that have "meaning" for it. One of his numerous examples is the tick:

> The constitution of the tick, which is blind and deaf, is composed only to the end of allowing every mammal to appear in its environment as the same carrier of meaning. One can describe this carrier of meaning as a radically simplified mammal, which has neither the visible nor the audible properties by which species of mammal are differentiated from each other. This carrier of meaning for the tick has only one smell, which comes from mammals' perspiration and is common to all. Furthermore, this carrier of meaning is touchable and warm and can be bored into for blood withdrawal. In this way, it is possible to find a common denominator for all these animals, so different in shape, color, emission of noises, and scent expression, such as we have before us in our environment. The properties of this common denominator fill in contrapuntally and activate the vital rule of the tick at the approach of any mammal, be it human, dog, deer, or mouse.[9]

More generally, he claims:

> For, in the environment of animals, every carrier of meaning is utilized through perception and effectuation. In every functional cycle, the same perception-effect process is repeated. Indeed, one can speak of functional cycles as meaning cycles whose task is determined to be the utilization of carriers of meaning.[10]

While, unlike Fodor, Uexküll understands the animal's world as a world of meaning, his main concern is to differentiate animal worlds from human worlds, as his book title indicates. Hence he presents the tick's *umwelt* as including a cut-down version of the human category "mammal." My concern is rather to explore the genesis of significance, of meaning, in the first place. My amoeba is not cutting and pasting features that are meaningful within the human *umwelt*—or within the scientific, physical world. It constitutes its own world made up of objects-for-amoeba that match the capacities of the amoeba. Meaning is not an exterior vernier painted over a pre-existing object that exists in-itself; the significance of food-for-amoeba is what constitutes it as an object in the first place.

The organism constitutes its own niche, its own world-for-the-organism, but this is not to claim that it produces it materially. Constitution is not production: the amoeba does not bring some new material into existence. Rather, it classifies matter already in the world into a new class. In unifying its detectors for these various substances by connecting them to a single approach mechanism, the amoeba has conferred a unified meaning on the various elements that make up its food. It is not the production of new stuff; what is constituted is a new meaning.

The constitution of meaning is central to Husserl's phenomenology, so should we label my analysis "phenomenological"? Husserl's philosophy is founded on the phenomenological "reduction:" one can describe what is given in experience only when reference to how things are in-themselves to "reality"—is held in abeyance. Within the reduction, he finds that consciousness experiences objects—what he calls intentionality—and that such objects are constituted by subjectivity as units of meaning (*Sinn*). Note that "reduction" here has nothing to do with reductionist explanation of everything on a physical or material basis. Quite the opposite: Husserl's methodological "reduction" effectively insulates the world of meaning (the phenomena) from causal processes that scientific and philosophical theories attribute to things-in-themselves and frees us to describe such meanings solely as they are experienced (phenomenology). Within the reduction there is nothing but meaning.

While all meaning, he holds, is constituted by subjects, these are not psychological or empirical subjects but special constituting ("transcendental") subjects. Indeed, in the case of cultural and scientific objects whose meanings are shared in common by many people, constituting subjectivity involves intersubjectivity. Of the many kinds of worlds consciousness constitutes (mathematical, scientific, fictional, and so on), the most fundamental world on which all the others are based is the everyday lifeworld (*Lebenswelt*): "to take them [realities of every kind and level] as existing in the absolute sense is consequently a countersense. … all real unities are 'unities of sense.' Unities of sense presuppose … a sense-bestowing consciousness. … An absolute reality is just as valid as a round square."[11]

Tønnessen proposes that we extend Husserl's concept of lifeworld and apply it analogously to the worlds of animals.[12] The *umwelten* would then be their "phenomenal" worlds (as opposed to the world of physics) and be made up of objects insofar as they appear to, are given to, or are experienced by the organism. Tønnessen suggests the term Uexküllian "phenomenology" for this project. Husserl himself occasionally hints at such a possibility: he says that the cultural lifeworld is referred "in its own way, to the human being as person, i.e. to the life of the soul." But then he adds, "And all this can be applied by analogy—just as far as the analogy reaches—to animals, to animal society, to the surrounding world with its specifically animal signification."[13]

While this phenomenological way of putting things is tempting, I have two objections. First, I worry about the terms appearance (phenomenon), consciousness, and subjectivity. Analogies and metaphors risk offering explanations whose plausibility depends on dragging in associated concepts inappropriate to the field toward which they are targeted. To speak of an amoeba's phenomenal experience successfully distinguishes its world from purely physical processes, but I fear it succeeds by covertly attributing consciousness to amoeba, an attribution I find absurd.

One response to this worry might be to split off the concept of conscious experience from the concept of subjectivity, and associate the constitution of significance only with the latter. In a way, Husserl's reference to intersubjectivity is a step in this direction in that it hints at a nonconscious social constitution. More fruitful, however, is Merleau-Ponty's notion of a body-subject. Discussing the human situation, he claims that the motricity of the "lived body" constitutes not only the unity of the body, but simultaneously the objects of perception and the space in which they are found. He rejects "intellectualism" and sees conscious thought as a later development that occurs only after the constitution of the lifeworld has already been accomplished by the subjectivity of the body. Against Husserl, meaning is not only for consciousness, it is initially generated by a "subjectivity" of the body.

According to Merleau-Ponty, "Bodily experience forces us to acknowledge an imposition of meaning which is not the work of a universal constituting consciousness, a meaning which clings to certain contents. My body is that meaningful core which behaves like a general function, and which nevertheless exists …"[14] In this modified sense of subjectivity that excludes consciousness, my amoeba might be said to be a (body-)subject whose motricity gives significance to its world.

I have a second objection to the metaphor of an organism experiencing a phenomenal world; however, that will allow me to develop my own analysis further. Even with the restricted sense of "body-subject," the metaphor sets up a dichotomy: the organism is an active constituting subject while the objects in its world are passive in the sense that they are the recipients of meaning donated by the organism. Organisms constitute; objects are constituted. This is an organism-centered account of constitution that still smacks of Husserl's subject-centered account of constitution. Let me now move beyond this way of describing the world of the amoeba to an account based on system. My toy organism-centered approach has given us insight into the nature of worlds of significance, yet its organism-centeredness is a serious limitation. To explain this, and to get beyond it, let me move up a couple of billion years and use a different scenario: that of bees and flowers.

Constituting system

The perceptual system of bees has ultra-violet detectors. They use these detectors to track down ultra-violet-colored flowers for their nectar. Bees pick out ultra-violet-colored flowers as a class and respond in a similar way to all of them—by flying to them and sucking their nectar. As I've been putting it, ultra-violet-colored flowers are objects constituted in the world-for-bees—the bees' *umwelt*—by the ultra-violet detectors of the bees and the responses that follow. For the bees, these flowers are pragmatically categorized by the normative motor value they have: they are to-be-flown-to; that is the general "significance" these flowers have in common.

From one perspective (that of the bees!), we can offer a simple adaptationist explanation as we did in the case of the amoeba: given the presence of ultra-violet-colored flowers in their environment, bees with the best ultra-violet-detectors have a selective advantage. Over a period of generations, most surviving bees will be the ones with better ultra-violet detectors. The evolutionary history of bees has created a world-for-bees, a world whose categories allow these flowers to be as objects. Since it is bees equipped with these detectors that categorize the flowers, we have an organism-centered perspective on the constitution of ultra-violet-flowers-for-bees.

But there is another perspective on the situation: the flowers'. To reproduce, flowers need insects to pollinate them and, in an environment with ultra-violet-perceiving bees, those flowers that have the best ultra-violet pigments have a selection advantage. Over time, more of the surviving flowers will have ultra-violet pigments. From this perspective, the flower is playing the role of (body-) subject, of constituter, and of giver of significance. Thus, the flower categorizes the bee as an object in its environment.

Both of these alternative perspectives are based on a simplistic interpretation of evolution: individual species are selected for their fitness within a given or static environment. But the environment is not static; it too evolves! A more sophisticated account—"co-evolution"—proposes that what evolves is not the bees or the flowers, but the interdependent system of their interaction.

Ulanowicz labels the process "autocatalytic mutualism."[15] This term brings out the dynamic, positive feedback aspect of the relationship. It is not just that bees rely on flowers and flowers on bees, the adaptive development of each provokes further adaptive development in the other. What we have here is not a matter of two perspectives, but of a mutual "enaction" as Varela calls it. Discussing the bee–flower case, Varela et al. put it this way:

> This insistence on the codetermination of mutual specification of organism and environment should not be confused with the more commonplace view that different perceiving organisms simply have different perspectives on the world. This view continues to treat the world as pregiven; it simply allows that this pregiven world can be viewed from a variety of vantage points. The point we are making, however, is fundamentally different. We are claiming that organism and environment are mutually enfolded in multiple ways, and so what constitutes the world of a given organism is enacted by that organism's history of structural coupling.[16]

Since the flowers also co-evolve with organisms in the soil that deliver nutrients to their roots and which in turn feed off decaying flowers in the soil, and since the bees have their own co-evolvers, there is a complex mutual relationship in which each component receives positive feedback from many other components. The simplistic explanation based on a dichotomy between a privileged active, adapting organism and its passive environment—of categorizer and categorized—must be superseded by a wider interaction of many active components that together give each other significance.

From an evolutionary point of view, we can say that it is not the bees themselves that evolve; nor is it just the flowers and soil. Rather we have a dynamic set of relationships, a system that evolves as a whole. Tønnessen speaks of "*umwelt* transitions" and claims that "What evolves is the way of life."[17] It is this historically developing system that constitutes unities of significance within it. It is not just that flowers-for-bees are functional objects in the bees' world or bees-for-flowers in the flowers' world: both kinds of objects are defined by their functional role within the ecosystem as a whole. While it is true that bees have perceptual organs, it is not these organs that are at the origin of the categorization of the flowers. Being an ultra-violet perceptual organ itself is a functionally defined entity within the overall system. This is no longer a subject-centered or organism-centered account: it is the interactions of the system as a whole that constitute both the organisms and their world of objects. Organisms are as much unities of significance as are objects. Which entity counts as an object—bee or flower—is relative to whatever one temporarily picks as the focus organism—flower or bee.

This systemic account gets us beyond the metaphors derived from Husserlian consciousness, Merleau-Ponty's body-subject, or Uexküllian phenomenology. It also takes us beyond the organism-centered toy world of my amoeba. It is the self-organizing evolutionary system as a whole that assigns various functional roles to its components. Organisms, their perceptual and motor organs, objects-for-organisms, food, poison, predator, prey, and so on are all components that make sense only in the context of the overarching system.

Nevertheless, the systemic account preserves some major features that the amoeba scenario allowed us to recognize. First, while naturalistic, the account remains non-reductionist: objects within a system are meaning-objects. To understand how the components of the system interact we must look not to their mechanistic properties, but to their functional roles. Different flowers may use different pigments for their ultra-violet color, that is, pigments governed by different chemical and physical laws. But this is irrelevant for understanding which objects the system constitutes. Similarly, which physical processes are involved in the bees' capacity to detect ultra-violet light makes no difference to our account. The processes are pragmatically defined: any process that can carry out the same function will do. The organic system therefore has a formative role in which the parts interdefine each other with a certain independence from the physical laws that may subvene. Kant already anticipated this point:

> An organized being is, therefore, not a mere machine. For a machine has solely *motive* power, whereas an organized being possesses inherent *formative* power, and such, moreover, as it can impart to material devoid of it—material which it organizes. This, therefore, is a self-propagating formative power, which cannot be explained by the capacity of movement alone, that is to say, by mechanism.[18]

The position I'm defending is not reductionist, but neither is it idealism based on subjectivity. The Western history of dualism can lead to a misleading dichotomy of matter and subjectivity that is exclusive, so that what is not one must be the other. The order of organic life, I have been arguing, is neither. Bees and flowers, amoebas and their objects, like all organic systems, are emergent structures that cannot be understood on the basis of physics alone. But that doesn't imply that they must be subjective or mental. Unification and significance originate not from matter, nor from subjects or organisms, but from the evolving system itself.

Selfhood

Introduction

Let us now move up another few hundred million years in evolution to the human species. My claim is that the kind of analysis I have offered for amoebas and bees can, by analogy, also give us an account of human selfhood. The world that a self lives in is also characterized by the features of unity, generality, significance, function, history, and so on. Selfhood is novel and distinctive, but it builds on and remains in continuity with the structures I have outlined for organisms. In particular, I claim that the self lives in a world-for-the-self—a human *umwelt* in Uexküll's terms—in which perceived objects and unities of action are functionally defined in parallel with the capacities of the self. This parallelism is a manifestation of a larger system—the social, cultural, and linguistic system—that constitutes both the self and its world.

In this very brief sketch of what such an account would be like, I will discuss, first, a caricatured "tradition" as a foil to bring out the distinctiveness of my account. Secondly, I will present the world-for-the-self as composed of cultural objects and a repertoire of actions. Thirdly, I will claim that selves and the world-for-selves are both constituted in parallel by the cultural system.

Traditional view

Let me first contrast my position with a certain common-sense view of the self as a self-subsistent entity. The Aristotelian view is that each human being is a substance defined by its own indwelling essence. For Descartes, an individual mind is a spiritual substance distinct from the body and independent of history, society, and language. Combined with religious notions of an immortal soul, these ways of thought merge into the Early Modern, "traditional" view of the self as an independent, self-sufficient, and self-conscious mind that secondarily perceives an independent world of reality, of objects-in-themselves. These worldly objects include other people who must be negotiated with to set up a social, linguistic, and political order, as Hobbes suggests. In the face of mechanistic science, the approach holds that the only way to preserve the dignity of the self is to insist on the impossibility of explaining it by physical processes—of "reductionism." This traditional position sets up a dichotomy: either one embraces self-matter dualism, or else declares that selves are an illusion caused by physical laws. The analysis I have given for biological life, when applied to human life, allows us to go through the horns of this dilemma.

Cultural umwelt

A person, like any organism, perceives biological objects, such as food, but is also capable of perceiving cultural objects. A human can see a cake as food-for-humans, but may also perceive it as a celebratory object that is part of a wedding ceremony. The wedding is a pragmatically defined object in that it fulfills a need for the individual, but this need, unlike the amoeba's, is a culturally defined need. The wedding is a constituted object in the sense that the various physical movements involved are unified into the one celebration and categorized in the same class as many other sets of movements that count in cultural practice as a wedding. Similarly, depending on the culture, a group may be perceived as a "political party," pieces of metal as "money," a mechanical structure as a "car," institutions as "universities," and so on. Just as in the amoeba case, these objects are general in that they are categorized and repeatable, and, since they are multiply realizable by many different material structures, they are defined by their cultural meaning, not by physics. This is an ontological point. It is not as if there were an in-itself collection of movements to which a human culture secondarily attributes the meaning "wedding." It is the cultural meaning that constitutes the object in the first place.

Analogous to the way that biological functions are generated by the history of the species, the meanings of cultural objects are generated by the history of the particular culture. Both biological functions and cultural patterns do more than describe facts; they set up norms. They establish what *should* count as food, or the way a wedding cake *should* be. Just as amoebas need unified detectors to detect food-for-amoebas, so an essential factor in becoming a self is developing the perceptual capacities to see the world as made up of culturally meaningful objects. From the physical or biological perspective, cultural objects are invisible. Perceptual capacities and perceived objects go hand-in-hand: perceiving and treating pieces of metal as "money" is what constitutes them as money. Since money is a cultural object, the perceptual capacity correlated with it is a learned activity. The self as a perceiver of money is as much a cultural construct as the object, money, itself.

It is similar in the case of action. Voting, jousting, making marriage vows, stealing, being faithful, going to war, and so on, are actions, rather than physical movements, only because of the intention of a self as agent. But the self is not sovereign, as the "tradition" would have it, for the capacity of a self to form such an intention derives secondarily from the culture. It is the culture that determines the repertoire of possible actions. In a culture in which the institution of private property does not exist, an individual cannot "steal." But, in a correlated fashion, the individual self must "fit" its world. The culture doesn't just constitute stealing; it also forms the kind of self that has the capacity to perform the act of stealing. Cultural learning constitutes a self capable of such actions.

Self and its system

Apart from learning particular actions, a self comes to learn what it is to act in the first place. A unified amoeba is, in a very weak sense, an agent that can be said to act as a whole. In the cultural context, however, an action is that for which the self can be held responsible. A knee-reflex may be a biological unity, but it is not an action attributable to a self. Responsibility is at the very core of selfhood and distinguishes it from an organic agent. I unify myself as the same person over time by taking responsibility today for yesterday's actions, promises, assertions, and beliefs. Claiming ownership of them, viewing them as mine, is the form of my enduring selfhood over time.

In learning what an action is, in learning what it is to be responsible for one's behavior, one learns to be a self. In the traditional dualist account, the subject is an ultimate given. Even when, as in Kant or in Husserl, the subject is not a substance, it remains a privileged, autonomous, self-sustaining entity that is the origin of meanings in its world. Just as the organism-centered account I rejected for the bees and flowers failed to see the organism as itself constituted by the ecosystem, so the ontological privilege granted to the subject in the traditional account neglects the historically generated, cultural nature of selfhood. An individual self comes to be within a culture. A baby may start life as an organic being, but its parents soon praise and blame it for its behavior; they foist responsibility onto it. What are initially organic movements come to be viewed by the child as hers, as her own. By such means, the child learns what a unified action is, what it is to be a responsible agent, what it is to be an enduring self. This is not just a matter of knowledge, of learning an intellectual concept: it is the process of actually becoming a self in practice. A self is a cultural construct passed on by learning from generation to generation.

The self is unified by this cultural construction. The self is not a physical or biological unit: reflexes, homeostatic balances, growth, and so on may be part of the human organism, but they do not count as elements of selfhood, for the self has no responsibility for them. Behavior when sleepwalking, when drugged, or when legally insane are not components of the self's unity. One organic human being may, in the case of multiple personality disorder, involve more than one self, with various actions attributed to one or other self. It is what counts culturally as an action, it is what society attributes responsibility for, that is integrated into one self. Selves do not come prefabricated; they get their unity and their being from the cultural and historical system within which they live. In giving meaning to actions and percepts, this system simultaneously, in the same move, sets up the self as a unified structure with the capacities to perform these actions and to recognize these percepts. Just as bees and flowers co-evolve in a mutual relationship, selves and their objects are constituted by the same historical impetus.

The ontological subordination of the self to the cultural system should not be misunderstood as stripping the self of its dignity or freedom, as relegating it to the role of a puppet or robot controlled by society. Our culture constitutes the self precisely with the functional role of a responsible entity. Behaviors that are attributed to the self are free actions that can succeed or fail, that can be praised or blamed. Like all functional systems, culture is shot through with normativity. It sets standards for what will count as an action in the first place, for what will count as this particular kind of action, for which actions are appropriate in the circumstances, for which are to be judged useful or useless, legal or illegal, morally right or wrong. Even beliefs are judged by a norm: rationality. Selfhood itself is normative. It is a standard, a standard the criminally insane, for instance, do not meet: insofar as they are not acting freely, they have fallen below the norm that the system defines as selfhood.

To summarize, this account presents selves as analogous to amoebas and bees on a number of counts. All live in worlds defined by pragmatic significance or meaning, as opposed to the physically defined world. Each involves a system that constitutes the organism or self as correlated with or fitting the world it lives in, as two sides of the same coin. Both biological functions and cultural meanings are fundamentally normative.

Conclusion

In this chapter, I have offered an integrated analysis of biological life and human selfhood. The analysis is naturalistic in the sense that it calls on no principles beyond nature. It is not, however, reductionist, for it appeals to modes of explanation other than physical and chemical laws. Both organisms and selves are novel entities that emerge out of processes we can understand by examining very simplified, just-so, evolutionary stories.

The central concept is the notion of constitution, the giving of significance or meaning. The notion is borrowed from idealists, especially Husserl, but I have naturalized it by stripping it of its association with consciousness and subjectivity and showing how the very simplest organisms live in a world of pragmatic, functional significances. This is, I claim, the most essential feature of organic life and allows us to distinguish it from mechanistic processes without appealing to any non-natural principle. The significances that characterize the organic world are constituted by the evolved system within which the organism itself comes to be. The notion of system permits us to understand constitution without attributing it to organisms as if they had some kind of proto-subjectivity. Finally, I have sketched very briefly how the same analysis could be applied to selfhood without appealing to a sovereign, absolute, or constituting subject.

The result is an approach that integrates human selfhood into the organic and physical worlds while maintaining the meaning and normativity that characterize and dignify human life, indeed all life.

Notes

1 From the Ancient Greek root πρᾶγμα (pragma), "a thing done, a fact." Related to πρᾶξις (praxis) an action. http://greeklexicon.org/lexicon/strongs/4234/
2 Merleau-Ponty 2003, p. 171.
3 Brandom 1994, pp. 86–7. The embedded quotation is from Hegel, *Phenomenology of Spirit*, para. 109.
4 Some, such as Sterelny, distinguish features of the world that are behavior-specific in this way from later evolved "decoupled" representations of intersensorial objects. Only the latter would count as fully-fledged intentional objects. For the purposes of my simplified toy world, I am leaving aside such complications, valid as they are. (Sterelny 2003, pp. 29–50.)
5 Kant 2007, p. 205 (5:377, para. 66).
6 Merleau-Ponty 1963, p. 156.
7 Fodor and Piattelli-Palmarini 2010, p. 140.
8 Fodor and Piattelli-Palmarini 2010, p. 143.
9 von Uexküll 2010, p. 179. I assume his "vital rule of the tick at the approach [of any mammal]" is conceptually similar to my Amoeba Food-approach Law.
10 von Uexküll 2010, p. 150.

11 Husserl 1982, pp. 128–9.
12 Tønnessen 2011 (= Analecta Husserliana CX/110), p. 331.
13 Husserl 1970, p. 228.
14 Merleau-Ponty 2005, p. 170.
15 Ulanowicz 2009, p. 76.
16 Varela et al. 1991, p. 202.
17 Tønnessen 2009, p. 48.
18 Kant 2007, p. 202 (5:374, para. 65).

References

Brandom, R. (1994), *Making It Explicit: Reasoning, Representing, and Discursive Commitment*. Cambridge, MA: Harvard University Press.

Fodor, J., and Piattelli-Palmarini, M. (2010), *What Darwin Got Wrong*. New York: Farrar, Straus and Giroux.

Husserl, E. (1970), *The Crisis of European Sciences and Transcendental Phenomenology*, trans. David Carr. Evanston, IL: Northwestern University Press.

—(1982), *Ideas Pertaining to a Pure Phenomenology and to a Phenomenological Philosophy, First Book*. The Hague: Martinus Nijhoff.

Kant, I. (2007), *Critique of Judgement*. Oxford: Oxford University Press.

Merleau-Ponty, M. (1963), *The Structure of Behaviour*, trans. Alden L. Fisher. London: Methuen and Co.

—(2003), *Nature: Course Notes from the Collège de France*, trans. Robert Vallier. Evanston, IL: Northwestern University Press.

—(2005), *Phenomenology of Perception*, trans. Colin Smith. New York: Routledge.

Sterelny, K. (2003), *Thought in a Hostile World: The Evolution of Human Cognition*. Oxford: Blackwell.

Tønnessen, M. (2009), "Umwelt Transitions: Uexküll and Environmental Change," *Biosemiotics* 2(1): 47–64.

—(2011), "Semiotics of being and Uexküllian phenomenology," in *Phenomenology/ Ontopoiesis Retrieving Geo-Cosmic Horizons of Antiquity*, vol. 10, ed. Anna-Teresa Tymieniecka. Dordrecht: Springer Verlag, pp. 327–340.

Ulanovicz, R. E. (2009), *A Third Window: Natural Life beyond Newton and Darwin*. West Conshohocken, PA: Templeton Foundation Press.

Varela, F., Thompson, E., and Rosch, E. (1991), *The Embodied Mind: Cognitive Science and Human Experience*. Cambridge MA: MIT Press.

von Uexküll, J. (2010), *Foray into the Worlds of Animals and Humans : With a Theory of Meaning*, trans. Joseph D. O'Neil. Minneapolis, MN: University of Minnesota Press.

11

Is Experience Subjective or Objective, or Both, or Neither?

John R. Shook

Questioning experience's subjectivity or objectivity sounds familiar, yet somehow outdated. Scientific psychology (and sociology) no longer rely on the concept of "experience" for much more than gesturing at what they really theorize about, or tentatively labeling preliminary issues for introductory textbooks. To the extent that the sciences have broken up the vague notion of experience into matters more amenable to separate inquiry, "experience" is receding in academic significance, returning to its folk psychological status where people converse about their mundane concerns. To the social sciences and humanities, things like conduct, conversations, and texts are plenty enough to study. It may be natural enough to talk about one's subjective experiences, but looking for the objective place of "experience" within nature may be quite unnatural. Is experience compatible with nature, much less an integral aspect to nature?

If we cannot answer affirmatively in light of our best science, then loose talk about "naturalizing" experience or consciousness, or establishing an "empirical naturalism," is just hot air. Pragmatists have typically thought not only that experience is just as "natural" as anything science already studies, but also that features and modes of experience somehow reveal crucial traits of nature itself.[1] Pragmatism takes the sciences the most seriously of any philosophy, yet science is supposedly concerned only for understanding nature's ways objectively, irrespective of whether a subject happens to be gazing on. Only inquiring subjects do science through their experience, it can be admitted, but the status of scientific knowledge about reality must not depend on whether any subject also confirms that reality in their personal experience. Philosophy has already asked and answered the question of what happens when subjective experience stands as the final judge of reality. (It is not anything like science and bears little resemblance to naturalism.) No, experience is endlessly interesting to us, but that is not because it possesses its own objective natural reality. After all, if it could be so independent, we would not so much care for it. We care because we experience, and we experience because we care.

In ordinary ways of talking, saying "that was an experience!" has emotional significance, perhaps as a heartfelt sign of appreciation or a cathartic outburst of relief. Promising an uplifting experience, or an amazing experience, or a comforting experience, similarly signals expectations about what will be encountered in the course of living. People have those encounters and undergo those experiences, individually or collectively, in the natural course of their days. (If you do not think people can share an experience, try paying more attention while attending a music concert or while gathering around the television to witness breaking news.) Experiences are expected to at least have affective influences on people, with greater (or lesser) intensity and impact, and some sort of lingering consequence. Perhaps sensory impacts never registered by anything in the nervous system, or transmitted up the nerves but filtered out by preliminary cortical processing, or retained but then ignored as "noise" and unused by higher cognitive processing, could somehow still count as "experience" in some sense that strains its plain meaning. But we must not start off by supposing that "experience" has a sharp meaning—consult the *Oxford English Dictionary*'s lengthy entry about "experience" to disabuse yourself of that notion. But the basic sense of "experience" is multi-sided and often vague, for good reason—its multi-purpose utility for plain talk by plain folk depends on such versatility. The social sciences and humanities deftly navigate that versatility, inquiring into peoples' experiences just by noting how they conduct themselves, how they describe what's been happening to them, how they learn from engagements with goings-on around them, and how they recount their stories about themselves, others, and wider events. What people have been doing, undergoing, learning, and narrating can largely exhaust the significance of "experience" for most inquiries into what it is like to be part of humanity.

The sciences are obligated to eventually inquire into anything and everything of potential relevance. Experience itself could not stay unexplored by the life sciences and the natural sciences forever. Psychologists (and cognitive scientists etc.) would not ordinarily say that their research has nothing to do with experience—quite the contrary. Very little holds more interest for us than the dealings of experience. Yet "experience" is no longer treated by the sciences, if it ever was, as a natural kind possessing its own inherent characters to be passively tracked and described for its own sake. If experience has been treated anywhere as a "natural kind" possessing its own inherent mode of being, that place is philosophy. The cognitive and brain sciences say that they study experience, but they craft, as they must, selected and specialized aspects to what has been considered experience. Some things that once were called "experiences" have been set aside by the mind/brain sciences as unreal, having no status at all, instead being granted a demoted status as illusions. Where is that odd experience, so worrisome to medieval people, of feeling impotent from demonic possession?[2] (The social sciences can make their inquiries into what people used to *say* about feeling impotent, but that singular experience itself has "vanished" as unreal because demons never existed.) These cognitive and brain sciences have also discerned fields of sentient "awareness" operating "behind" or "below" conscious experience, so that the meaning of "experience" has undergone growth by addition, as well as subtraction.[3] The reasonable theory that sufficiently complex animals possess some level of sentient "awareness," and perhaps an inner field of phenomenal consciousness as well, has further extended the meaning of "experience" beyond its folk psychological sense.

Discriminable features to the living processes of organic life usually deserve their own labels or even scientific terms in the course of close study. So much has been discriminated and aggregated about the nervous processes of animal life that continuing to expect any single quality, feature, function, or structure to continue to be found in every instance of an "experience" is no longer a mark of scientific psychology or sociology, or biology in general. But taking note of this development is just another way of pointing out how scientific psychology departs from folk psychology and ordinary language. Experience, by itself, may be a poor guide to what is really going on with experiences.

Perhaps what brings together what is going on with experience, if not any inner unity or evident essence, has more to do with the life forms having experiences and those organisms' ecological habitats. There's a reason why we expect experiences to at least be closely associated with animals. Experience should not be nonbiological, in any radical sense, and the biological always has a home, a habitat. Failure to situate experience and sentience where life is ongoing can inspire speculations about unnatural habitats. Philosophers from Anaxagoras and Plato down to Descartes and Kant have not left those metaphysical options unexplored. What must the sciences do? Explaining the sentient animation of complex animals can be treated as an empirical question instead, as Aristotle and Darwin suggest. At this point, where science takes a divergent path from metaphysics, four very puzzling issues about experience emerge.

First, if experience is (supposedly) the place from which all empirical evidence could arise, how can there be "evidence" of experience? Trying to discover evidence about experience may be either uninformatively circular, or uselessly paradoxical. After all, any evidence must already have been within experience to be informative, so one cannot produce satisfactory "evidence" without simply producing some experiences directly. Some phases of experience can serve as pointers to other phases (by functioning as signs, for example); but there is no "evidence" about experience that is not just more experience. Put another way, it is paradoxical to think that there is something that serves as evidence of experience that is not just comprised entirely of experience itself. (Try to relieve someone's thirst with only "evidence" of water but not any water.)

Second, even if a distinction could be intelligibly widened between experience and some evidence of experience, that gap cannot exceed the bounds of experience itself. Everywhere our sentience goes, we always find some experience ongoing for us. Experience may not be reducible without remainder to sentience (some extraordinary experiences or "subconscious" experiences may lack sentience), but sentience without any experience in the broad sense would not be happening. Intelligible evidence will never be available to detect where "experience" stops at its necessary limits, for we cannot "see" the other side of the boundary line, if there even is one. Only our capacities for experiencing time, motion, directionality, and perspective, along with recallable and comparable episodic memory, permit us to fully realize that our experience is indeed bounded. A being that somehow experienced without temporal or spatial displacement, and never grasped direction and perspective, would take its field of experience as unbounded, and it would probably never acquire the concept of "experience," either. (As a thought experiment, do not think about a deity, but just consider a newborn baby.) Furthermore, we acquire concepts of individual bounded objects in our dynamic reaches of experience, but no amount of experience will ever display a "perspective" to experience as a whole. Where would you have to "stand" to gain a second perspective on all of experience as a whole? We speak of gaining perspective from experience, but we do not actually think that experience itself is the thing we are acquiring perspectives on—we are only using an elliptical way to talk about different specific experiences, or perspectives on a prominent matter within experience.

Third, even if we could formulate a method to assign natural boundaries to "experience," and hence manage to "localize" stretches of experience within some wider realm that is not just more experience, the genuine sciences are forbidden from relying on unnatural entities, for either evidentiary or explanatory roles. Why would a delimited amount of "experience" be unnatural? Our experience is characterized by noticeable directionality and perspective, so time and space are automatically involved, but that is hardly the same thing as knowing that any delimited stretch of experience itself possesses its own definite temporal and spatial dimensions from moment to moment. Science is not so atomistic anymore, so let us not demand that a "piece" of experience has a narrowly specifiable location in space and time. All the same, it may be a category mistake to treat a stretch of experience as an "object," an "event," or even a fluctuating "field" amenable to scientific study. If you (rightly) expect that quantum mechanics cannot say where an electron's momentum is precisely located at a particular moment, then you should appreciate how psychology would not appreciate being expected to say much about the location, dimension, energy, and trajectory of a stretch of experience, much less any of its other physical properties (e.g. tell me the mass of your experience of the soccer match). For all practical purposes, no science will be experimenting directly with some "units" of experience or any aggregates of experiences, since those "things" are not really things. Ultimately, it is very hard to understand how some delimited stretch of experience could count as observably objective evidence for anyone but the experiencer, or how stretches of experience could be responsibly postulated as theoretical "entities" since they lack minimally natural properties.

Fourth, even if we could scientifically manipulate experiences as respectably "natural" things for confirmable experimental results, one more feature to experience in general is its teleological character. Common sense says, and cognitive science confirms, that what experiences an organism is having are largely a function of what the organism happens to be seeking. We only contingently find what we began to seek, but what we eventually find had something to do with our pursuits. This is valid in both a mundane sense and a psychological sense. For common sense, while we sometimes feel that events come to find us (especially unwanted events), honest reflection tells us that we also had to end up there to receive the "gift." Wisdom also says that gifts show up right in front of us from time to time, yet we do not recognize many of them because we are minding other things. As for psychology, it has been established almost as a truism that what we notice has more to do with what our cognition was guiding us toward at the time than what is simply impacting our bodies to be aware of. In a crucial sense, what is most interesting about experience is not how it happens when external things affect the body now, but instead why experiences exist only for some future end. It is impossible to fully account for experience without postulating something about the future for which ongoing experience mainly exists and is primarily attempting to become. The ends somehow "produce" the causes. But this no longer sounds very scientific. Postulating future (not yet existing) ends and attributing to them causal powers in the present, or, going even further, postulating real guiding powers already existing in the future to manipulate present materials toward future ends, is a manner of "explanation" deemed unacceptable and untouchable by empirical science. Setting aside as completely illegitimate either sort of "teleological" or "design" type of explanation has been widely regarded as necessary for policing the boundaries of acceptable empirical inquiry and hence science itself.[4]

The boundary between unscientific metaphysics and empirical science may be thick and fuzzy, but these four deep puzzles about experience show why it must be thrown into the arms of one, or the other. Unless all four puzzles can be resolved in favor of permitting the biological and cognitive sciences to effectively inquire into "experience," science must never speak of it with any ontological seriousness, nor can it ever be "naturalized" as genuinely real. Experience would then be left to the plain meanings of folk psychology and ordinary language, where the social sciences and humanities can do their proper work studying people, not "experience." Experience would also continue to have other friends too. The imaginative musings of natural theology and the untethered speculations of pure metaphysics are always happy to welcome experience to an unnatural home. Alliances between anthropology and theology, say, or sociology and metaphysical dualism, as another example, are theoretically possible, and various options have been explored already. Empirical science can turn its back on experience, but it then loses intellectual authority to pass decrees against the misuse of experience.

Pragmatism can get entangled with these "unnatural" alliances. Let anthropology accept without disqualification the religious experiences and mythic narratives taken as real by devout peoples, and a natural theology can then "discern" the work of the divine in the human realm. The resulting pragmatic confirmation for the divine (how could something so humanly functional be just unreal?) would not be contradicted by those "harder" sciences—the life, cognitive, and natural sciences—which have nothing to say about experience. Similarly, so as those harder sciences stay silent, an alliance between sociology and metaphysics can describe phases of experience as noetically transcendent, reaching depths (heights?) of unnatural realities through the "right" sorts of social training for the practicing faithful. Scientistic societies do not encourage these experiential voyages, so their disdain toward the transcendent is both predictable and irrelevant to the experimental methods "confirming" the reality of transcendent realms. Both sorts of alliances have received endorsements from numerous nineteenth- and twentieth-century intellectuals who rely on vaguely or explicitly "pragmatic" methodologies. In fact, some of these figures, whether they regarded themselves primarily as theologians, philosophers, psychologists, sociologists, anthropologists, or just humanists, have also claimed alliances with naturalism.[5] Not all pragmatic naturalisms are the same, evidently. If confident reliance on experiential methods of wide utility for humanity is sufficient for affirming naturalism, which I for one deny, then both of these anthropological and sociological pragmatisms will expand the realm of the "natural" into transcendent places where naturalism was never designed to go. Naturalism was never supposed to be about taking as genuinely real what people happen to traditionally or experientially prefer to believe is real. The question comes down to, what sort of "science" shall control naturalism?

Naturally enough, both of these alliances are well aware of the inevitable resistance from scientific naturalism, which only accepts as scientific the theorizing backed by exploratory experiments into what nature is doing, and not just what people are saying. That is why both alliances put great effort into diminishing the authority of the harder sciences, by depicting those sciences as not just practical but optional methodologies, meriting no top-ranking right to decide what is exclusively real over other useful modes of human experiencing. Lacking any right to dictate what is real, what people talk about returns to a level playing field. On that field of equality, the "subjective" experience of an individual could be just as "valid" as anything "objective" asserted by a scientific field, or at least immune from invalidation by any scientific worldview. Or, perhaps the collective experiences of a society agreed on a description of reality could be as "objectively" valid as, or at least not disprovable by, anything asserted by a scientific tribe mired in their own group-think "subjectivity." If the harder sciences want to reply how they are not so impressed by any and all human "experiencing," then they had better back up that judgment with good reasons. If the harder sciences are going to manage to say something intelligible about experience, and especially about what counts as more or less informative experiences, then they had better be able to experimentally theorize about experience, at least on their own terms. So the struggle comes down to this: how can "experience" be somehow "placed" within nature so that it can be subjected to exploratory experiments by some of the "harder" sciences? And here we have returned again to the four urgent puzzles about experience.

Having observed what is at stake for both pragmatism and naturalism, we can set down *the "postulate" of "experiential naturalism*:" "experience" is amenable to scientific inquiry if those four puzzles are resolvable in a jointly coherent way. That resolution will set the parameters for guaranteeing that experience itself is naturalizable in some thinly provisional sense as a legitimate "object" of scientific inquiry. Now, experience may turn out upon thorough multidisciplinary examination to be natural only in some thinly derivative or epiphenomenal sense, akin to the way that "shadows" are natural enough events but lack an ontological status in their own right, or the way that "wetness" is natural only in an epiphenomenal way by lacking its own efficacious powers, since properties of water molecules are actually at work. Both shadows and wetness can be experimentally investigated by the sciences because the four deep puzzles about "locating" them in nature are resolvable, and evidence about them is elicitable. If the postulate of experiential naturalism is workable and scientific inquiry into experience proceeds, we may further expect that the manner of resolving those four puzzles about nature's "place" for experience is already providing fundamental information crucial to the next great theoretical problem about experience: since experience seems to be naturally existing, how and why does it exist?

We may predict that great progress can be made toward confirming what can be labeled as *the "postulate" of "naturalistic empiricism:"* the results of scientific inquiry into experience will yield an impressive confirmation for the independent ontological theory that experience specifically has an ecological home with sentient organism–environment situations. Experiences turn out to be right where they naïvely seem to be: where sentient organisms are engaging their local environs. In other words, experience intuitively appears as directly encompassing the external objects and events we can and do interact with, and naturalistic empiricism confirms how this is no hallucination or mass illusion. We are not forced to accept any skeptical notion that experience is mostly or entirely deceptive, or any metaphysical notion that we can only experience some "inner" or "transcendent" realm.

Finally, confirmation for naturalistic empiricism in turn permits confidence in *the "postulate" of "pragmatic naturalism*:" our ordinary methods of intelligently exploring our environment typically yield good evidence about natural events, and after suitable refinement and extension, accumulated evidence can methodically yield reliable knowledge about reality's ways. No speculative philosophical work is needed to ponder whether "experience" ever accurately "mirrors" the external world, nor is any epistemological theorizing needed for adding a priori assurances that experiential evidence carries enough "truthfulness" to in turn justify theorizing about the world. Pragmatist philosophy of science does have to explain how sophisticated experimental inquiries can justify the credibility awarded to the amazing theoretical discoveries of modern science, but that explanation will only display continuities with our ordinary empirical methods rather than any dramatic chasms. Pragmatic naturalism is therefore in a soundly scientific position to adjudicate those cultural struggles over anthropology–theology alliances or the sociology–metaphysics alliances. Pragmatic naturalism balances the judgments of the social and life sciences concerning what seems real in experience with the authority of the natural sciences on what actually can be experienced. (Classical pragmatism, for its own part, never encouraged such rankings, nor the "unity of science" movement in philosophy.) All sciences must have workable alliances with each other, and renounce other rival alliances. Integrating, without subjugating, all of the sciences in a single worldview is the ultimate ontological aim of pragmatic naturalism, with natural theology and transcendent metaphysics reasonably excluded as lacking any grip on reality.

Grand philosophical strategies must humbly wait upon the provisional "naturalization" of experience. That naturalization rests on satisfactorily resolving the four puzzles. For our modest purposes here, we shall proceed through the four puzzles in reverse order, from the most specific to the most general. This procedure is scientifically warranted, since what the cognitive and life sciences can say about experience is piecemeal but concrete.

The fourth puzzle noted how experience is highly teleological: its biological status is one of purposefulness, in a double sense. What is experienced has mostly to do with what needs to be experienced by the organism in the course of its activities, and experience itself in some sense gets created and designed in service for the goals of those activities. Not only are the contents of experience featured for attaining goals, experience itself as a whole is created "by design" for guiding the organic processes attaining those goals. Even our "passive" and "subconscious" awareness of so much affecting us is proactively filtered and purposefully re-processed before rising to any level of conscious awareness, much less due attention. In short, not only does what we are doing largely control what we *can* attend to, our behaviors are controlling what we *do* attend to, and hence what we think is really happening around us.[6]

Our cognitive systems are primarily sensitive to only what they habitually take to be most relevant to the ongoing behaviors, including alarming shocks and distracting interruptions. Incoming information about what is happening is blended together with pre-set anticipations of what should be happening (what those behaviors are supposed to be accomplishing), so that blended information can be mostly directed toward ongoing feedback loops—including "feed-forward" loops—for dynamic motor control at the subconscious level. What also gets passed along toward conscious awareness consists only of heavily blended, edited, redacted, and strongly exaggerated versions of just the most relevant matters before us. (That moment-by-moment "updating" creates the difference between "focus" and "fringe," and also gets manifested as what Dennett has described as the "multiple drafts" feature to consciousness.)[7]

Our cognitive processes literally create what we attend to and monitor, as we move our bodies around, enact our purposes, and engage with our environments. That is the point of enlarged nervous systems and bulbous brains—primitive fish developed them and surface animals relied on them, as habitats grew ever more complex and surprising. Many animal species took the evolutionary path growing their cognitive capacities. An incremental improvement in cognition turned a bit more of the surroundings into the smarter animal's "environs" (whatever can be relevant to behaviors). That enlarged dimensionality to the environs brought more contingent instabilities, calling for additional expanded cognitive capacities, which further enlarged the environs of possibilities, and so forth. There are natural limits, of course. Each species' evolution usually reaches a metabolic equilibrium, unable to grow larger brains due to constraints on energy consumption. A species typically remains in its environmental niche, and hence its environs stays the same, until further speciation (or extinction) occurs. Primates successfully managing quite dynamic environs in Africa due to changing environmental circumstances grew quite large brains, and a few *Homo* species took advantage of those brains to dramatically complicate their social environs as well, sparking a further "brains race" that produced our species.[8] Our species not only has an incredibly wide ecological "environs" that people can navigate, but we also have to think about our social "emplacement" as well.[9]

The point to this evolutionary tale is that the experienced environs for an organism is not identical to its entire environment. Brain cognition teleologically creates informative experiences primarily about what should be encountered on a moment-to-moment basis, rather than statically registering everything impacting the organism that just happens to mechanically occur. We can now suggest a resolution to the fourth puzzle. The fourth puzzle contrasts the teleological nature of experience—what "is" in experience is mostly about what "should" be—against the scientific demand that teleological explanations shall be excluded from methodological consideration. It appears to be impossible to account for experience biologically without postulating something about the future for which ongoing experience mainly exists and is primarily attempting to become. Are we really talking about the future ends somehow producing the present causes?

Fortunately, the sketch of a cognitive account of experience sketched here does not involve a role for some actual "existing" future. Nothing in the future has to "exist," much less be psychically detectible or magically efficacious on the present. Experience is biologically created for the sake of the future in a vaguely anticipated sense, but not for any specific "existent" future which must exist in some sense. All brain cognition needs to do is habitually function to anticipate particular patterns (and deviations from those patterns) of detectible consequences that result from ongoing behaviors. Those cognitions habitually function that way because those patterns have been typically detected in the past, not because cognition somehow knows how they will occur in the future. Anticipation is not prescience, predestination, or providence. Cognition is not sneaking a peek into the future, noting what shall be the case, and then adjusting nervous processes to re-direct conduct. The only efficacious relationships to all this cognition are acceptably scientific ones from the present toward the future. Ongoing experience helps create the near future, at least future situations within and nearby the organism. But this is just common sense. Cognition simply takes advantages of sensory modalities attuned to pre-anticipated sensory results from behavior in order to quickly process them for modulating ongoing motor activity directed toward some envisioned goal (like relieving thirst, eluding a predator, and so on). That continually reactive-and-adaptive process is primarily affective and emotional, rather than purely intellectual. When present in awareness, this forward-directed "attending to something" constitutes what can be labeled as "intentionality."[10]

Resolving the fourth puzzle justifies a good measure of credibility for "empiricist naturalism," which experience will prove to be amenable to scientific inquiry. Can the third puzzle be resolved now? That puzzles questioned whether any delimited stretch of experience itself possesses its own definite temporal and spatial dimensions from moment to moment. If not, then no science will be experimenting directly with any units of experience or any aggregates of experiences, since those "things" are not really things. If so, then experience can be assigned some spatial and temporal dimensions, not of its own apart and away from nature, but within nature itself. (Again, if the "dimensions" to experience never coincide with nature's, then experience might only have a transcendentally metaphysical reality but surely not one amenable to direct scientific inquiry.) To resolve this puzzle, some intelligible sense must be given to the notion of an experience possessing some assignable temporal and spatial dimensions within nature, so that a science can know where it is and how to interact with it. What do we know about "where" and "when" experience exists?

Experience can be a proper subset of the environment, since the cognitive sciences confirm that it must be subject to three constraints. First, the experienced environs is only a selected portion of the whole environment—the environment is always larger than the range of experience because the sensitivity of organic sensory modalities (hearing, seeing, smelling, etc.) has natural limitations. Second, the experienced environs at any given time has a focal area where sensory intensity maximizes due to physical proximity and hence mechanical efficacy. We see things best when they are nearer (but not too near). We taste things best when the tongue gets a thorough opportunity. We hear things best at close range. We smell things best when our noses can sample large amounts of air passing by. But this quantitative sense of "closer" is already dependent on our qualitative appreciation for intensity, not the other way around (a baby is not judging the distance of its rattle toy first, before estimating the loudness of its rattle). Third, we learn in our infancy that moving our bodies correlates well with adjusting the intensity of sensations, and also with the coordination of multiple senses agreeing about nearby objects (that rattler is loud, round, red, reachable, and lickable). We learn the notion of "here"—where sensations are most intense—as opposed to "there," where things are most interesting. Our experience therefore always has two "focal points:" the place where I am (which I can move around) and where interesting things are (which have multiple perspectives). Brain cognition accomplishes this all, designing ongoing experience to select out what in the environment shall be present, and intensely interesting in the moment, for the organism.

Because of the way that experience has those two focal places, one "here" and the other "there," the "subject" (me) and the "object" (the object of attention) are cognized through organic nervous processes so that both are within the environment and the environs. (It must be strenuously asserted that this convenient labeling of one focal point as "me" is not the same as the separate, and false, claim that any experience whatsoever is subjectively experienced by a unified self, an ego, an "I," or anything like a self-conscious center. What it is like to be a mouse includes one focal point where the mouse is, and the mouse's cognition tracks where it is, but there is nothing like a "subjective me" or "I" for that mouse.) This simultaneous presence within both environment and environs is easily proven to be the case. First, the experienced "me" is simultaneously within both the environs and the environment because a finite body that I call "me" occupies the "here" place, which, so far as I or anyone else can tell, is continually part of the environment and also within other peoples' environs (when they are near me). If this "subject" was not part of the environment and hence exempt from physical constraints and laws, and unreliably detectible in other peoples' environs, then "my" experience would be profoundly different from what it actually is. (This thought experiment might be good for imagining what being a ghost would be like.) Second, the experienced "object" is simultaneously within the environment and my environs because (1) if it was only in the environment but not in my environs then I would not experience it, but we are only here speaking of experienced objects; and (2) if it was only in my environs but not in the environment then nothing else in the environment could directly causally effect it, since things in the environment can presumably only causally affect matters in their physical proximity.

Situation (2) does not rule out the possibility that I can experience some things only in my environs but not really in the environment; it only rules out the possibility that absolutely everything in experience is only in my environs but never in my environment. The difference between an object in *both* my environment and my environs, and an object *only* in my environs, is easily discriminable. I can first check whether other objects which are securely in my environment can directly make any change to this curious "object." If so, then it is an external thing too. If none of them can do so, then I can check whether I can, by moving my body, cause a change to it or arouse a different perspective on it. If so, then it is oddly "close" to me or within my body but no hallucination. Only after exhausting those circumstances do I start to judge that I am experiencing something having nothing to do with my environment at the moment but only my inner states (like a perceptual malfunction or a cognitive error arousing a hallucination or a delusion). Things really in my environs/environment do not display all of their causal efficacies, traits, features, and so forth all at once—it is the real things that are only experienced as real partially and perspectivally, and it is precisely the unreal (only in experience) things that have uniform and non-perspectival features. Returning to the main point, if my experience really consisted solely of my created environs and never my external environment, then I would assume that all of reality consisted only of what I experience, which has been philosophically labeled as "solipsism." In summary, I am neither a ghost nor a solipsist.

Our brains are active architects of designed experience. That sounds like experience cannot be the same thing as the environment, even if it in some sense "recreates" the environment. Quite the opposite is the case, as they are ontologically one, although the experienced environs is always a smaller portion of the environment. An experienced environs is part of the environment the way that an ocean wave *is* part of the ocean, and not like the way that the movie studio's constructed set of a battleship's bridge "is" a part of a battleship.

Let us first deal with the claim that the experienced environs is always a smaller portion of the environment. We know this for two reasons.

First, from experience we learn that more environment is beyond the current environs because every time we move to check, more environment emerges into experience, letting us realize that the environs has a "horizon" of finitude, even if we cannot in one sitting simultaneously enjoy an environs and also perceive that environs from the other "side" of the boundary at the horizon. There are always "incoming" environmental matters causing eruptions of events and causing objects to change and move within our environs.

Second, on the imagined hypothesis that our environment is always absolutely identical to the environs we perceive, we would not possess our conception of the causal powers of objects. Suppose I see a squirrel come into view outside my window. Not even imagining that it "came" from beyond my view (for, by hypothesis, I cannot think anything really exists there), I must either accept its spontaneous creation or credit something happening within view for causing the squirrel to abruptly exist, but I cannot credit the squirrel with its own powers of animation while not in view. Suppose I next see tree branches outside my closed window, but not the entire tree, and those branches are moving in an irregular fashion. Unable to think that there is more "tree" beyond my view and unable to think about "air" beyond my view, I cannot credit anything invisible with causal efficacy making the branches move. If I had a conception of causality, it would be an extraordinarily magical one, where the wildest correlations can constitute "causation." However, we large-brained primates do not use that thin notion of causation (if it is even worthy of calling "causation"), but instead we regard it as cognitive error. We have a much richer and refined understanding of causation and the causal powers of objects, precisely because we never think that our environs exhaust the environment. Indeed, animals as "lowly" as fishes and crustaceans probably "get" that difference at some minimally sentient level. Furthermore, even lowly animals enjoy that overlap between their environs and their environment. It is not a question of cognitive firepower. Once cognition rises to the level of taking advantage of the detectible difference between environs and environment, ever-smarter animals continue to rely on it, and so do we. We can even remember the many noticeable differences between the ever-wider environment and our delimited environs thanks to our short-term and long-term memories. On this second point, we know that objects having their own stable causal powers can enter our environs and cause events within our environs.

Having established that the environs is always “smaller” than our environment—yet matters going on in the wider environment continually arrive into the environs and causally affect matters in the environs—the main issue remaining for the third puzzle is the question of ontological identity. The claim made here by empiricist naturalism is that our environs simply is our environment, taken in partially rather than wholly. That is to say, for example, when an object naturally existing in the physical environment enters an organism’s environs, there is no secondary ghostly “duplicate” or phantasm of that object that actually arises in that environs—the object itself is experienced. There has always been two primary objections made against that identity thesis: one metaphysical, and one scientific.

First, it is metaphysically objected that the natural object cannot be identical with the experienced object, on the grounds that two things A and B can be the same singular thing only if every feature/trait/power/localization/etc. of A is the same as those of B. Put another way, if there is anything different between A and B at a moment in time, besides perspectival/relational properties, then A cannot be identical with B (readers may apply their favored version of the traditional principle of identity of indiscernibles). Applying this point to the claim that the experienced object in the environs is identical with the real object in the environment, this claim is reduced to the view that the experienced object must possess all of the real object’s intrinsic properties. Unfortunately for empiricist naturalism, this metaphysical reasoning judges, it cannot claim that this is true. Objects as experienced rarely display all of their real properties at a glance while within view in experience. Therefore, empiricist naturalism is false, according to this metaphysics. What happens to appear within experience, it cannot be ontologically identified with any real object existing in the environment. Experienced objects can be “of” the natural world but never “in” the natural world. So says metaphysics.

How can empiricist naturalism reply? The logical answer is obvious: the principle of identity of indiscernibles is being misused. Empiricist naturalism holds that it is the real things that are only experienced as real partially and perspectivally, and it is precisely the unreal (only in experience) things that have uniform and non-perspectival features. Applying the principle correctly, empiricist naturalism judges that the real object in experience possesses its own additional properties, over and above any available as experienced, in order to explain how the object passes into, and out of, the environs of experience without any catastrophic change or any ghostly double emanating from it. The metaphysical argument fallaciously substituted one claim for another.

Empiricist naturalism does not deny that *the experienced object* must possess all of the real object's intrinsic properties, and it is not also committed to the (false) view that *the object as experienced* must possess all of the real object's intrinsic properties. Therefore, this metaphysical argument fails to show that the experienced object cannot be identical to the real object in the environment.

Second, there is a scientific objection, proceeding from empiricist naturalism's view that experience is dynamically created by the bio-chemical processes of the nervous system's cognitions. If the experienced object really exists in the environment, and also in the environs as experienced (they are not two ontologically distinct things), then the organic cognitions within an organism are creating external objects in the environment. Could it really be that the house I see across the street is supposedly created moment-by-moment by my own cognitions? Science will never find that theory remotely plausible. Even metaphysics and theology would hesitate. If I were a solipsist with miraculous causal powers, perhaps that notion could become plausible, or perhaps I am Bishop Berkeley's God. However, if empiricist naturalism is contrary to that divine solipsism, then it must stop claiming that experience is creating external real objects from moment to moment. Metaphysics could take this opportunity to re-suggest that cognition is simply creating a ghostly "double" of aspects to external objects, with some causal input due to impacts on the senses from the (never experienced) external affairs.[11]

It does no good for empiricist naturalism to bitterly complain that this metaphysical suggestion leaves the external environment in a transcendental condition, or that this metaphysics cannot explain what "input" would specifically arrive, or how it could "impact" us, from that mysterious realm. Empiricist naturalism is even more ridiculous if it must affirm that cognitions within us must create the objects outside us—that the "me" is entirely responsible for the "object." Empiricist naturalism itself claims that the "me" and the "object" are separate matters within environs/experience, but that must be illusory since the object is thoroughly dependent on the "me" in some deeper way. In fact, since the "brain" is doing the cognizing yet this brain is itself an "object" that people can observe (with effort), then the brain itself is created by ... cognition? Cognitive science itself collapses, since it cannot be true that nervous systems are doing the cognitions, so there must be transcendental cognitions first miraculously creating nervous systems, which then somehow create experiences, which finally(!) create external worlds. This all seems like a Rube Goldberg contraption, even for a transcendental deity.

All the work undertaken to make empiricist naturalism plausible has collapsed. We have reached the point of the second puzzle: what is evidence, and what is experience? The very distinction has collapsed. All that is left is a metaphysical distinction between evidence/experience and something transcendentally beyond all possible experience. We can still say that we possess concepts about individually-bounded objects within our dynamic reaches of experience. But no amount of experience will ever display a "perspective" to experience as a whole. You'd have to "stand" with the transcendental to gain a second perspective on all of experience as a whole in order to really explain it.

Before resigning ourselves to the transcendent one way or the other, a preliminary premise put into use in preceding paragraphs should be re-examined. Empiricist naturalism holds that the organic cognitions within an organism are creating external objects in the environment. That sounds bizarre, but it must be interpreted rightly in light of another crucial view of empiricist naturalism: only the real things (not the entirely experiential things) are experienced *as* real partially and perspectivally. Assemble these points and think them through together: organic cognitions are creating experienced things partially and perspectivally. Now set aside the metaphysical complaint yet again, that those experienced things cannot be ontologically identical with objects in the environment. What is actually experienced is those environing things, partially and perspectivally. We have reached the claim, which empiricist naturalism must make, that cognition is involved with environing things displaying real (but partial and perspectival) features, and these features would not really be features of those objects unless some cognitive processes were involved.

We have reinvented the affordance-capacity distinction theorized by J. J. Gibson.[12] A feature of an object is an affordance if it is creatively cognized in the course of an organism's capacity to interact with that object. The affordance is a stable feature of the object where it can be repeatedly and reliably cognized as such in the course of the organism's successful engagements with it. Those repetitive engagements make the affordance a feature of the object—unless those engagements were repeatable and successful, so the organism's cognitions learned to adequately anticipate an object's affordance, no affordance would arise. Different sensory modalities are connected to different sorts of affordances, and many affordances are the result of two coordinated modalities. Coordinated sensory modalities especially permit us to figure out when an affordance will function adequately, or perhaps inadequately. For example, I know that the hammer can be used to strike that nail, but if I notice that the hammer head has come loose from the handle, I would not expect that affordance anymore. Furthermore, according to empiricist naturalism, affordances never exhaust an object's real properties and powers. A real thing is always regarded as possessing its own causal efficacies while beyond experience, or while within experience. This is also easily confirmable through our experiences interacting with objects in novel ways to discover new affordances. Only cognitively guided interactions with objects yield discoveries of novel affordances, as we learn how to do original things with objects. If we instead supposed that the number of affordances to an object were forever fixed, or that an object's affordances could be created by sheer cognition without any interactions, we would never learn how to discover novel affordances to objects. Yet we surely do this, with immense satisfaction.

The way that objects have affordances is highly consistent not only with the demand that cognition participates in the establishment of objects' features, but also with the expectation that cognitive experiences of objects are teleological in nature. Because affordances are always paired with cognitive capacities, and those capacities guide ongoing bodily activity toward some goal, then the experienced affordances are *meaningfully* emphasized in experience. Put another way, there is purpose behind the way that the relevant affordances useful for ongoing activities are focally intensified in the organism's environs. Not that neuronal subsystems or entire cortical modules have purposes! No, all that is meant is that there is real purpose behind what gets emphasized in the environs and what does not: the organism's own purposes. The "subject"—the purposeful "me"—is responsible for the featured affordances of the "object" that are attentively experienced and used from moment to moment. It is this precise point which exposes representationalism as inadequate, since these meaningful matters in the environs cannot exist without the purposive organism, yet those meanings exist in the natural environs and cannot be discovered anywhere within the organism's cognitions or subjective awareness. The habitat of meanings is right where pragmatists have claimed it was all along, to the astonishment of dualist and subjectivist philosophers. Neither meanings, information, nor content is "in the head."[13] Cognition is also monitoring the "background" against which the attended object stands out but only for potentially disruptive or alarming shocks or patterns. That explains why the "fringe" background of broader experience feels far less intense and meaningful and gets entirely ignored during moments of deep concentration. Alternatively, if one can suspend all immediate purposes, cognition can relax the foreground–background distinction, so that one can have the feeling that nothing in experience is meaningful or that everything is equally meaningful (however one wishes to describe it).

The second puzzle can now be solved: there is no sharp ontological dichotomy between what is evidence and what is experience, but there is a real difference between focal matters of meaningful significance and background matters of near-meaninglessness. The sciences can therefore judge the approximate dimensions and focal points to a stretch of an organism's environs/experience by experimentally detecting what the organism is capable of finding meaningful, from a high level of intensity to a background level of bare sensitivity. This procedure would be impossible unless one or more affordances are accessible simultaneously by both the experimenter and the organism. Blowing a whistle in the vicinity of a dog will typically elicit some behavioral reaction, but blowing the whistle in the vicinity of a spider may never do so. Our immense array of affordances, including machine-enhanced affordances (we can be sensitive to a machine's sensitivities to natural goings-on) permit us to discern much of the range of an organism's experience/environs if we immerse ourselves into its "world" and manipulate its features. Remaining behaviors lacking any discernible affordances have to be attributed by abductive theorizing to undetectable natural features, but good theorizing can often suggest reasonable and confirmable answers. (Animal behaviors that turned out to be sensitive to tiny changes in magnetic fields took a long time to understand, for example.)

Cognitive and brain sciences do indirectly manipulate and experiment with delimited stretches of experience, of course, because sentient organisms are available for interactions. You do not need any fancy equipment, in fact: enter my visual and auditory ranges and ask me to shake your hand. When I then shake your hand, you might be thinking that you have successfully manipulated a stretch of experience other than yours (e.g. one of mine as well), even though you lack a solid idea about the real means (the "mechanics" involved with the causation) of doing that, and you are not situated exactly where I am to enjoy the exact same range of experience. Still, you can access some direct evidence for your success, because your experience shaking my hand does partially overlap in limited perspectival ways with my experience shaking your hand. I can know this, because we shook each other's hand. This happened because you had to regard my hand as a set of affordances just the way I do, and I had to regard your hand as a set of affordances just the way you do. Joint action on a prompting like that takes plenty of cognitive preparation and guidance—both of us already have to regard our hands in terms of ready affordances for grasping. Although the entire set of properties to my hand would not be in your environs, enough of its affordances will be attended to in your experience to guarantee that we are both experiencing the same hand, and you will simply take for granted that the rest of the hand's features are pretty much the same for both of us as well.

What will never be in my environs is your experience of the "internal" qualities of your hand; here, only abductive analogy lends reasonable clues, as I presume that your experience of what it like to feel your own hand is pretty much like mine with regard to my hand. As far as "external" affordances are concerned, the more we can coordinate our actions together in the same environment, the more we think (rightly) that our environs overlap to a greater and greater extent. Nevertheless, we retain our ability to distinguish our respective ranges of experiences.

The first puzzle now admits a resolution. Anything that serves as informative "evidence" about experience is just composed of more experience, it is true. One cannot produce satisfactory "evidence" without simply producing some experiences. This can appear to obstruct the possibility that I can gain evidence of your experience. For example, I can show you things on the table in front of us, which in some sense are "in" my experience. But if you demand that I quit showing you things in my experience and instead show you my experience of those things, I must either gesture at all of the things on the table again, or I must fall silent in failure. However, the evidence required to show how to experience what someone else is experiencing need not be the same as that person's entire experience. Similarly, the evidence required to show that another's range of experience is experimentally manipulable need not be entirely the same experience as what is being manipulated. Only affordances overlapping within environs are required, not any complete overlap to the entirety of the experience inquired into. There is no call for producing something to serve as evidence of experience that is not more experience, so this puzzle is side-stepped. For scientific inquiry into experience to commence, relevant evidence only has to coincide with a partial and perspectival portion of that experience.

My experience of the affordances of your hand shaking mine serves as evidence that I am modifying your experience of your hand in specifiable ways. The scientist's experience of slightly vibrating a spider's web at one string to make the spider scurry in the direction of that string serves as evidence that the scientist is modifying the spider's environs and (unconscious) experience in specifiable ways. Where the environs of a spider is located, and what the environs of a spider is (roughly) like, can be discerned experimentally to some degree. We will not know too much about what it is like to be a spider, but we can discover a little bit about what it is like to be in that spider's environs, which is not that unlike guessing at what it is like to be that spider.[14] The greater the cognitive capacities to an animal, the better we can be at gaining ideas about its environs. People who share their homes with cats often wonder what could possibly be going on in a cat's mind sometimes. But long, shared experience with a cat does reveal a great deal about what a cat takes to be in its environs (and what it ignores). The cat's experience will not be in our experience in all its inner qualitative splendor, but limited phases and ranges of its experience into its environs are accessible to us with long practice and due attention. Furthermore, because cat cognition naturally generates its experienced environs, which is an entirely natural process, the cat's experience is part of the natural world. Yes, there is something like what it is to be a cat, and it is not metaphysically locked away from the natural world. Neither is your experience, nor mine.

Having offered reasonably scientifically informed resolutions to the four puzzles about experience, we can defend *the postulate of "naturalistic empiricism:"* scientific inquiry into experience is yielding an impressive confirmation for the ontologically minimalist theory that experience specifically has an ecological home with sentient organism–environment situations. Experiences turn out to be right where they naïvely seem to be, where sentient organisms are engaging their local environs, without any resort to metaphysical speculations or postulated transcendentalisms.

The ultimate goal on this line of thought remains the postulate of "pragmatic naturalism:" our ordinary methods of intelligently exploring our environment typically yield good evidence about natural events, and after suitable refinement and extension, accumulated evidence can methodically yield reliable knowledge about reality's ways. According to pragmatic naturalism, the most reliable knowledge about reality is yielded by the experimental learning acquired though probing investigations into our experienced environment (directly or instrumentally), which requires the continual expansion of our explored environs.

Where do the social sciences stand? The social sciences do inquire into human–environs relationships and matters within the social world. The social world looms large for these fields, where the conduct and conversations of ordinary people are typically about that social environs, because that's where people can get things done with other people. The phenomenon of "social affordances" has been noticed but studied piecemeal and classified as other sorts of things: symbolic meanings of groups, social facts, public values, civic morals, and so on. Accounts that people relate about what is going on within their social environs are, reasonably enough, helpful for understanding how they experience that social environs and how they conduct themselves within it. The social environs is naturally real just like everything else that biologically cognitive humans do, but that does not mean that everything that people talk about is real just in the manner as they describe it.

Earlier I brought up the example of demonic influences. People talk about demons a great deal in some societies. Since the biological and natural sciences fail to confirm the existence of anything behaving the way that alleged demons do, then it follows that there are no demons and no one is experiencing what it is like to be infected by a demon. There are no real entities in the environment possessing the affordances of demons. However, there are phenomena within demon-infested societies which do display the social affordances permitting demonic powers to operate. The social sciences are in an excellent position to understand what it means to conduct oneself within a certain social environs which includes demonic powers capable of having relationships with humans. However, there will not be any anthropological or sociological inquiry into demons, but only into social environs where people talk with each other about demons. That is because no one, neither the social scientists nor the demon-believers themselves, undertake anything like experimental inquiries into demonic affordances to reliably confirm their predicted interactions with humans. Instead, "confirmations" of demonic influences by the most devout practitioners notoriously consist of routine perceptual and cognitive errors, confirmation biases, and placebo effects, to which we are all susceptible. The "demonic" affordances turn out to consist entirely of the deliberate manipulation of social affordances for sustaining "belief" by those unable or uninterested in starting intelligent inquiry, which usually includes ardent practitioners who sincerely believe too. It is unnecessary to be reminded that societies themselves take responsibility for classifying personal mental disorders, for the habitat of disturbed social cognition is also where sound social cognition lives, within societies themselves, as the viewpoint taken here is trying to assert.[15]

Social scientists must be overruled if some of them demand that demons be tolerated as somehow just as environmentally real as anything studied by the other sciences. That overruling is justified, not because any science or set of sciences has superior status over the rest, but rather because science comes first where effective reality is concerned, and those social scientists have no scientific inquiries into demons. People sincerely believing in demons are quite real and they must not be told that their sincere experiences are not real, but due perspective must not be abandoned. After all, any person's experienced environs is just a partial and perspectival portion of the whole surrounding environment, and anyone could be terribly ignorant about what is actually responsible for arousing one's cognitive capacities to shape their experiences the way that they do. Real matters out in the environment could be largely responsible, but cognitions make their contributions most firmly when a great deal is at stake, and there are many urgent matters in the social environs. People see many real things, but what they are really seeing has its own proper habitat, which may not be the external environment.

We literally "see" things that are not really objects or events in the environment, because it turns out that those things can be perceived only within our social environs. We can still tell the difference between those things that are really in both the environment and our social environs and those things that only exist in our social environs using tests similar to our intuitive tests for detecting "objective" things and "subjective" things. For example, we perceive the real value of specially printed pieces of colorful rectangular paper. But nothing that we securely regard as environmentally real can causally affect that paper's perceived value unless its printed symbols are modified or other symbols elsewhere are massively modified (such as destroying half of the currency supply or erasing all financial records at all banks). This fact about those pieces of paper tells us that we only perceive the meaningful value of pieces of paper because their printed symbols only exist for our social environs. Money does not exist in the environment, because its affordances are demonstrably not residing there. As they say, money doesn't grow on trees.

Pragmatic naturalism recognizes the intellectual capacity of each scientific field to accurately report on the natural matters to which they are individually attuned. Like individual brains, scientific communities sustaining their inquiry methodologies forge theoretical "environs" to be valid portions of enveloping nature. No single scientific environs will get it all right in all perspectives and dimensions. The best we can do is partially overlap them, prevent radical discontinuities, and avoid transcendent chasms. In the aggregate, the many sciences can forge a broad-spectrum worldview that tries to comprehend everything real, even though we suspect that there is always more left unknown.

Notes

1 The central work of pragmatism defending these stances is still Dewey 1929. The implications of these stances for epistemology and philosophy of science are explored in Shook 2000. Michael Eldredge emphasizes the cultural role of experience for Dewey's thought in *Transforming Experience: John Dewey's Cultural Instrumentalism*.

2 Consult, for example, Caciola 2003.

3 A survey of current neuroscientific thinking is assembled by Franklin and Baars 2010, pp. 91–102.

4 See Turner 2003, pp. 57–70.
5 William James is a seminal figure inspiring these entanglements and alliances. See Proudfoot 2013.
6 Consult the chapters of Calvo and Gomila 2008, and those of John Stewart et al. 2010. See also the analyses by Shapiro 2010.
7 See Dennett 1991.
8 See Sterelny 2003; Mitani et al. 2012.
9 Stewart and Strathern 2003, pp. 4–5.
10 See Okrent 2007.
11 On varieties of consciousness dualism, the reader can begin by consulting Robinson 2004.
12 Gibson 1979.
13 Hilary Putnam rightly credits the classical pragmatists for this claim about cognition in his own contribution to this radical notion in *The Threefold Cord: Mind, Body, and World*, 1999. Representations are real, of course, where humans craft them in order to behold them, but that is a specialized cultural technology. The anti-representationalism stance is elaborated in such recent works as Chemero 2009, and Hutto and Myin 2012.
14 Consult the fascinating discussion of spider experience in Jackson and Cross 2011, pp. 115–74.
15 Consult Klass 2003.

References

Caciola, N. (2003), *Discerning Spirits: Divine and Demonic Possession in the Middle Ages*. Ithaca, NY: Cornell University Press.

Calvo, P., and Gomila, T., eds (2008), *Handbook of Cognitive Science: An Embodied Approach*. San Diego, CA: Elsevier.

Chemero, A. (2009), *Radical Embodied Cognitive Science*. Cambridge, MA: MIT Press.

Dennett, D. (1991), *Consciousness Explained*. Boston, MA: Little, Brown.

Dewey, J. (1929), *Experience and Nature*, 2nd edn. New York: W. W. Norton & Co.

Eldredge, M. (1998), *Transforming Experience: John Dewey's Cultural Instrumentalism*. Nashville, TN: Vanderbilt University Press.

Franklin, S., and Baars, B. J. (2010),"Two Varieties of Unconscious Processes," in *New Horizons in the Neuroscience of Consciousness*, Elaine Perry et al. (eds). Amsterdam: John Benjamins, pp. 91–102.

Gibson, J. J. (1979), *The Ecological Approach to Visual Perception*. Boston, MA: Houghton-Mifflin.

Hutto, D. D., and Myin, E. (2012), *Radicalizing Enactivism: Basic Minds Without Content*. Cambridge, MA: MIT Press.

Jackson, R. R., and Cross, F. R. (2011), "Spider Cognition," *Advances in Insect Physiology* 41: 115–74.

Klass, M. (2003), *Mind over Mind: The Anthropology and Psychology of Spirit Possession*. Lanham, MD: Rowman and Littlefield.

Mitani, J. C., Call, J., Kappeler, P. M., Palombit, R. A., and Silk, J. B., eds (2012), *The Evolution of Primate Societies*. Chicago, IL: University of Chicago Press.

Okrent, M. (2007), *Rational Animals: The Teleological Roots of Intentionality*. Athens, OH: Ohio University Press.

Proudfoot, W., ed. (2013), *William James and a Science of Religions: Reexperiencing The Varieties of Religious Experience*. New York: Columbia University Press.

Putnam, H. (1999), *The Threefold Cord: Mind, Body, and World*. New York: Columbia University Press.

Robinson, W. S. (2004), *Understanding Phenomenal Consciousness*. Cambridge: Cambridge University Press.

Shapiro, L. (2010), *Embodied Cognition*. London and New York: Routledge.

Shook, J. R. (2000), *Dewey's Empirical Theory of Knowledge and Reality*. Nashville, TN: Vanderbilt University Press.

Sterelny, K. (2003), *Thought in a Hostile World: The Evolution of Human Cognition*. Malden, MA: Blackwell.

Stewart, J., Gapenne, O., and Di Paolo, E., eds (2010), *Enaction: Toward a New Paradigm for Cognitive Science*. Cambridge, MA: MIT Press.

Stewart, P. J., and Strathern, A. (2003), *Landscape, Memory and History: Anthropological Perspectives*. London: Pluto.

Turner, S. (2003), "Cause, Teleology, and Method" in *The Cambridge History of Science*, vol. 7, *The Modern Social Sciences*. Cambridge: Cambridge University Press, pp. 57–70.

Index

www.ingramcontent.com/pod-product-compliance
Ingram Content Group UK Ltd.
Pitfield, Milton Keynes, MK11 3LW, UK
UKHW020800290926
15470UKWH00003B/877

9 781472 511058